Bruce Buckley, Edward J. Hopkins, Richard Whitaker

weather

A VISUAL GUIDE

Published by The Reader's Digest Association Limited
London • New York • Sydney • Montreal

Contents

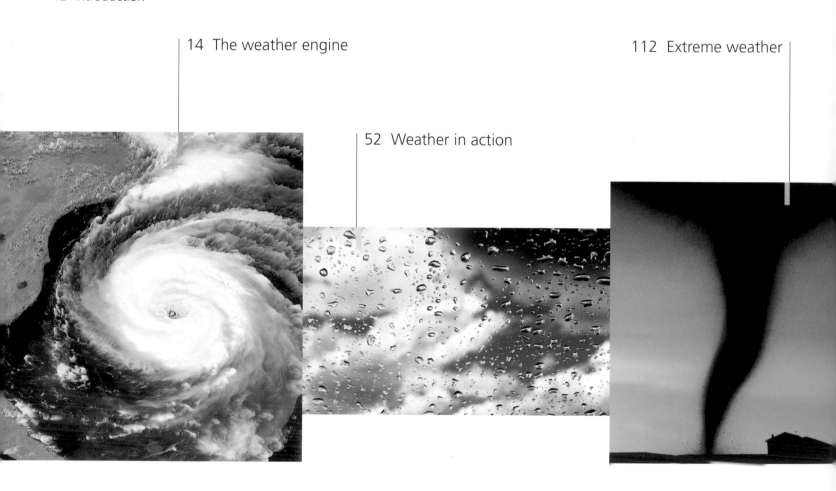

Introduction

As Mark Twain said, "The weather is always doing something." It is, indeed, a phenomenon of universal interest and fascination, as well as research and scholarship. The beauty of an unusual cloud, the symmetry of a rainbow, the awesome power of a thunderstorm and the devastation of a hurricane—all suggest a mighty power with influence over human affairs. Not surprisingly, early humans explained changes in weather as the actions of sky-gods who could reward or punish at will. Only in the last two centuries or so have scientific reasoning and experiment been used to explain the nature and causes of weather and to forecast its behavior. As this process developed, it was realized that because weather is a global phenomenon, international endeavor is required to understand and predict it. The international cooperation today provides one of the finest examples of international synergy united in a common cause.

But this book begins well before that. It traces long-term climate change over Earth's 4600 million-year history; explains the complex atmospheric forces that influence weather; examines the diversity of climate throughout the world and how plants, animals and people have adapted to it; analyzes the factors that interact to create violent weather extremes; and reviews the latest research into current climate change. This weather odyssey is one of the more remarkable chapters in the story of humanity—one that the reader will enjoy sharing.

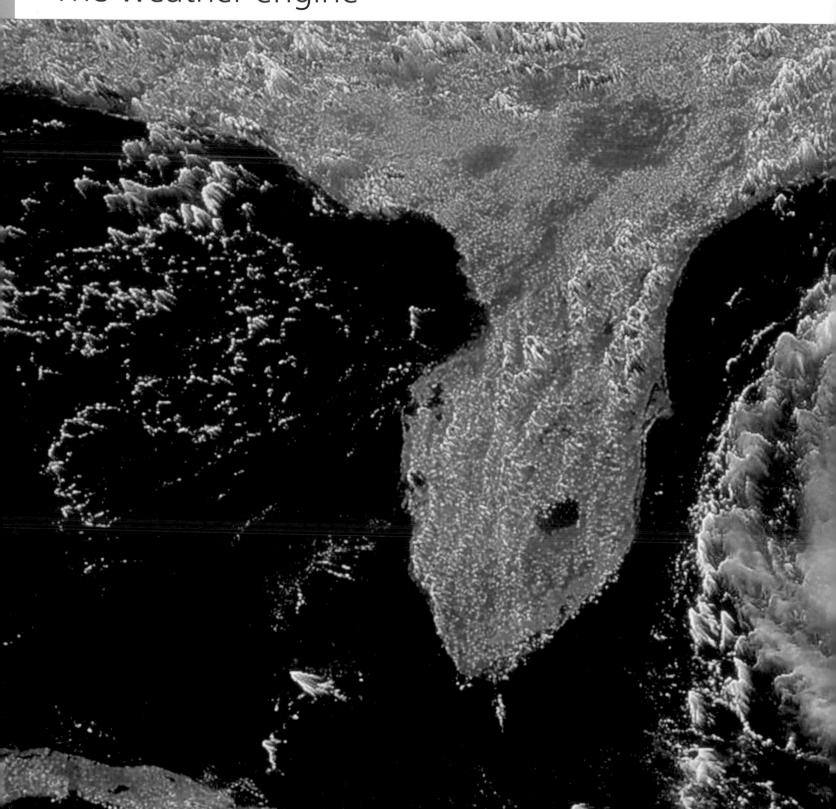

Previous page Hurricane Fran approaches
the coast of Florida in 1996.

The weather engine

Fueled by the heat of the Sun, the weather is a system of cycles and forces within the atmosphere, the blanket of air that surrounds our planet. Vast waves of air and swirls of cloud circulate in complex, everchanging patterns, giving rise to winds, storms and other weather phenomena.

What is weather?

Our planet, Earth, is surrounded by an envelope of air called the atmosphere. Weather is the state of the atmosphere at any particular time and location. It is probably humankind's most widely discussed topic, and its effects are all-pervasive, ranging from the trivial to the tragic. Weather dictates the kind of life we lead, our homes, our clothes, our leisure pursuits. The weather in any one region can vary from place to place and from day to day, or even hour to hour. Climate, on the other hand, is the typical weather for a locale and takes into account long-term averages and extremes. Minimally, accurate records for 30 years are needed in order to construct a profile of the climate of any given area. While meteorological records have been kept for only a couple of centuries, they can be supplemented by historical data and a growing body of evidence from the natural world to provide information about Earth's changing climate.

WHY WE STUDY WEATHER

One of the reasons people study weather is simply that they are inquisitive and wish to classify and explain atmospheric phenomena in simple terms. More important, however, is the need to anticipate the weather, so that we can prepare for extreme events or take advantage of favorable conditions. Meteorology—the scientific study of weather—is a relatively new discipline. Today, our understanding of the science of weather is reinforced by satellite images showing Earth from space. The photograph at right is a composite, created with data from three Earth-orbiting satellites. Hurricane Linda can be seen off the southwest coast of North America, and heavy clouds swirl around the polar regions. The Moon is an artistic embellishment.

↑ **A trick of the light** An optical phenomenon caused when sunlight is refracted by raindrops, rainbows are admired for their beauty. They can also be used to predict local weather in the short term.

↖ **The force of nature** Dark clouds can unleash dramatic storms featuring torrential rain, thunder and lightning, and damaging winds. Observing a storm is watching the raw power of nature. A storm's origins are complex yet, in many cases, predictable: a combination of atmospheric forces that together create the conditions that are unleashed in these intense phenomena.

← **Twisting winds** One of the weather's most violent aspects is the tornado, a funnel-shaped column of spinning air that can cause death and destruction. Tornadoes are the ultimate storms, their high winds capable of devastating any structure in their path and wreaking havoc on the landscape through which they pass.

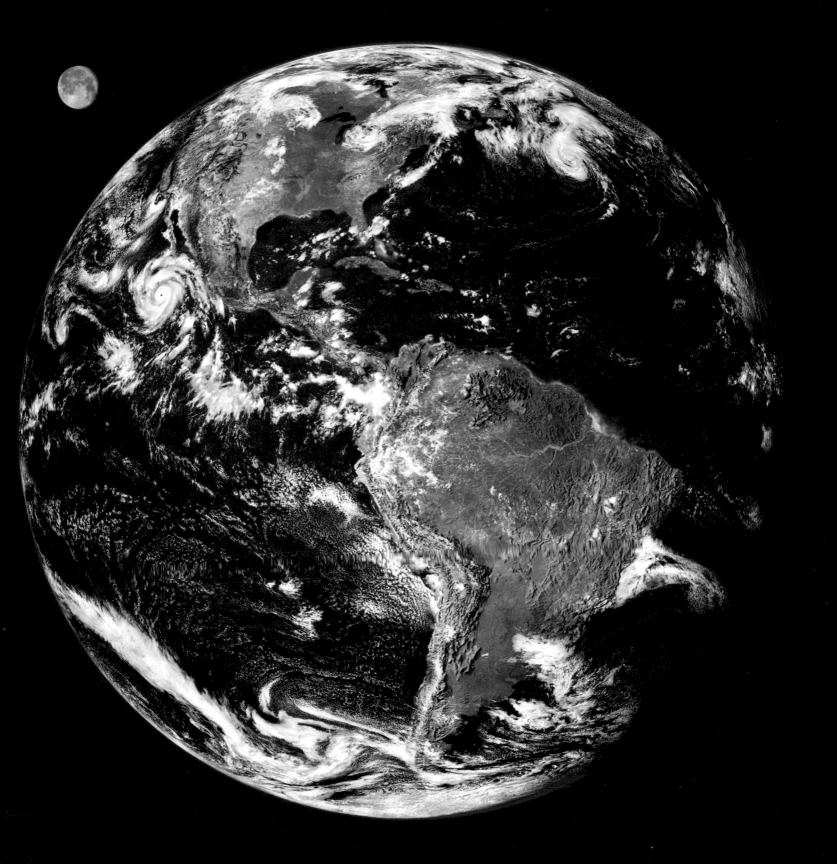

The solar powerhouse

The Sun, the ultimate source of nearly all energy on Earth, is a giant ball of gas that contains 99.9 percent of the Solar System's mass. A typical middle-aged star, the Sun shines because of the thermonuclear reactions that are taking place in its core. As these reactions fuse hydrogen atoms to make heavier helium atoms, they release enormous amounts of electromagnetic radiation that slowly travels to the surface. About 46 percent of the Sun's radiation is in the form of visible light, while an equal amount is near-infrared radiation, which we feel as heat. The remainder is ultraviolet radiation, the form that causes sunburn in humans. The Sun also emits a constant stream of high-energy particles as part of the solar wind.

SOLAR ERUPTIONS
Captured by the Extreme Ultraviolet Imaging Telescope, three hooked prominences erupted from the Sun's surface on January 11, 1988. Many times the size of Earth, these solar prominences are made of extremely hot plasma (about 125,000°F or 70,000°C) that has been bent and twisted by the Sun's powerful magnetic field. Another form of solar eruption, the solar flare, is a brief but powerful event that can release as much as 2 percent of the Sun's energy.

→ **Inside the Sun** Scientists use a technique called helioseismology to study the Sun's interior. Subtle pulsations at the solar surface resonate deep inside the Sun as sound waves. Measured by satellite telescopes, the pulsations can be graphed in a color-coded image that helps scientists infer the Sun's structure and composition. This increases our knowledge of the mechanisms operating from within the Sun.

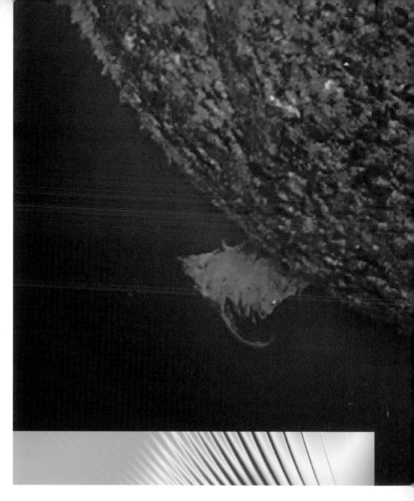

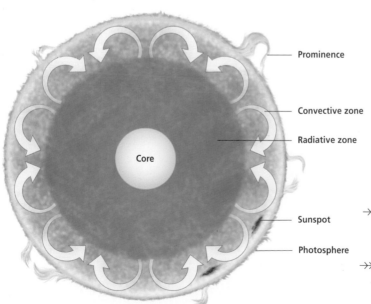

Prominence

Convective zone

Radiative zone

Core

Sunspot

Photosphere

THE STRUCTURE OF THE SUN
The Sun has a layered structure (*left*) Thermonuclear fusion reactions in the dense core produce temperatures of some 27,000,000°F (15,000,000°C). Energy from the core diffuses outward in the form of photons (parcels of electromagnetic energy) through the radiative zone. It then cools and undergoes a boiling, convective motion through the convective zone. By the time it reaches the photosphere—the Sun's visible surface—the gas is only about 11,000°F (6000°C). Here, cooler areas appear as dark sunspots, and loops of gas known as prominences erupt from the surface. The chromosphere, a thin, cool layer, surrounds the photosphere. The Sun's vast, hot outer atmosphere, the corona, has temperatures of 2,000,000°F (1,000,000°C).

THE SOLAR WIND
A stream of protons and electrons known as the solar wind flows into space from the Sun's corona at speeds of about 250 miles per second (400 km/s).

→ **Solar activity** The left half of the Sun in this composite image shows a coronal mass ejection, an eruption of gas often, but not always, associated with flares and prominences. Coronal mass ejections cause high-speed gusts in the solar wind.

→→ **Earth effects** The solar wind (in white) causes Earth's magnetosphere (blue lines) to become tear-shaped. Gusts in the wind can create geomagnetic disturbances, such as auroras, and even disrupt communications and power supply networks.

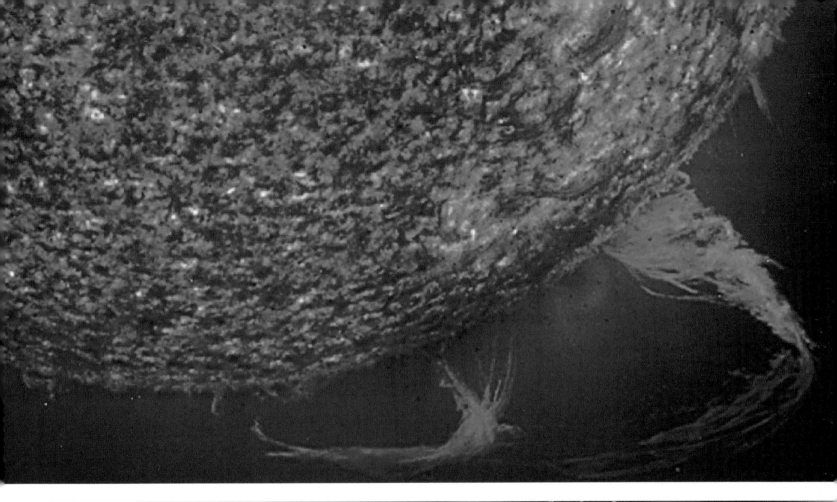

The seasons

The movement of planet Earth around the Sun produces daily and seasonal changes in the amount of sunlight, which result in cyclic changes in temperature. As Earth spins on its axis, it completes one rotation every 24 hours, creating night and day. Over the course of a year, Earth completes one orbit of the Sun. Because Earth's rotation axis is tilted at an angle of 23.5 degrees, different parts of the world receive varying amounts of sunlight at different times of year. The longest day in the northern hemisphere is the summer solstice, around June 21; this is the shortest day in the southern hemisphere.

REASONS FOR THE SEASONS

Earth is about 93 million miles (149 million km) from the Sun, and rotates around the Sun in an elliptical path, taking about 365 days to complete one orbit. Simultaneously, Earth rotates about its north–south axis in a counterclockwise direction. Earth's axis (the imaginary line running between the North and South poles) is not perpendicular to the plane of its orbit around the Sun. So, depending on the time of year, some latitudes are tilted toward the Sun, while others are tilted away from the Sun. As Earth travels around the Sun on its annual orbit, solar rays reach the planet at different angles. For half the year, the Sun's rays fall most directly on the southern hemisphere; for the other half, on the northern hemisphere. When Earth's axis is tilted away from the Sun around the December solstice, the northern hemisphere receives less sunlight and experiences winter, while the southern hemisphere has summer. At the June solstice, the situation is reversed. In temperate climate zones, this cycle leads to the four seasons of spring, summer, autumn and winter.

Equinox around March 21
Sun over equator: northern hemisphere spring, southern hemisphere autumn

Solstice around December 21
Sun over Tropic of Capricorn: northern hemisphere winter, southern hemisphere summer

Solstice around June 21
Sun over Tropic of Cancer: northern hemisphere summer, southern hemisphere winter

Equinox around September 21
Sun over equator: northern hemisphere autumn, southern hemisphere spring

↓ **Seasonal changes** Temperate regions exhibit great seasonal variations throughout the year. Temperature changes affect not only the appearance of the landscape but also the agricultural cycle of sowing and reaping. Equatorial regions are heated more consistently and experience little seasonal variation.

Polar day and night Locations near the poles receive about six months of almost constant sunlight, followed by six months of almost constant darkness.

Midlatitude seasons The Sun's varying path through the sky is responsible for the marked seasonal changes of midlatitude locations.

Tropical constancy In tropical regions around the equator, the length of day remains the same year-round, giving rise to a constantly warm climate.

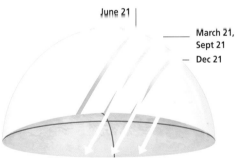

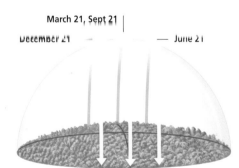

LOCAL SUN-PATH AT NORTH POLE

An observer at the North Pole sees the Sun make one complete circuit of the sky just above the horizon on the spring equinox in March, then attain its highest path in the sky on the June summer solstice. By the autumn equinox in September, the Sun is again just above the horizon. It then sinks and remains below the horizon until the next March equinox.

LOCAL SUN-PATH AT MIDLATITUDES

An observer at 45°N sees a marked seasonal variation in the daily path of the Sun. On the equinoxes, the Sun reaches a noon position of 45 degrees above the horizon and stays in the sky for about 12 hours. On the summer solstice, it travels higher and stays in the sky for more than 15 hours. On the winter solstice, it is lower in the sky and appears for less than nine hours.

LOCAL SUN-PATH AT EQUATOR

Along the equator, an observer would experience essentially no changes in the 12-hour length of daylight throughout the year. On the equinoxes of March and September, the Sun travels directly overhead at noon. On the June solstice, the Sun's path is displaced slightly to the north, while on the December solstice, the path is slightly to the south.

Earth's system

Our planet can be considered a system with five distinct components: atmosphere (air), hydrosphere (liquid water), cryosphere (ice), lithosphere (solid) and biosphere (life). These components are interrelated through the flow of energy and matter.

The flow of energy begins with incoming sunlight that warms the planet, ending with a spaceward flow of energy that prevents Earth from overheating. The incoming solar energy causes movement in the fluid atmosphere and hydrosphere. Some energy is used to drive the water cycle, in which water changes form, becoming solid, liquid or gas as it flows between the various components of Earth's system. Other chemical substances, such as oxygen and carbon dioxide, also cycle through the system's components.

Atmosphere A gaseous envelope surrounds Earth, protecting life from the harshness of space. The atmosphere's clouds, suspended particles and gas extend at least 60 miles (100 km) above the surface.

Hydrosphere The liquid water component of Earth's system consists of the oceans and other large bodies. The hydrosphere covers about 71 percent of the surface and contains most of the planet's water.

Cryosphere The component of the planet made up of ice is known as the cryosphere and includes glaciers and polar ice caps. A large proportion of the planet's freshwater is found in the cryosphere.

A GLOBAL SYSTEM

The interconnections of our planet's system components are evident in any image of Earth. An aerial view of Mount St. Helens, Washington, USA (*left*), shows snow from the atmosphere blanketing the volcano, which forms part of the lithosphere. In an image from Apollo 17 (*right*), we see the blue oceans of the hydrosphere, the Antarctic ice cap (part of the cryosphere), white clouds (atmosphere), continents (lithosphere), and dark areas of rain forest in equatorial Africa (biosphere).

Lithosphere The lithosphere is the solid portion of Earth, and includes the soil and rocks upon which we live. Nutrients from the atmosphere are fixed in the soil and used by plants in the biosphere.

Biosphere As far as is known, Earth is unique among planets because its system supports life. The biosphere is made up of animals, plants and other organisms, as well as decaying organic matter.

Biosphere interactions Green plants contain chlorophyll, which helps to convert solar energy into carbohydrate through photosynthesis. Oxygen is released into the atmosphere as a byproduct.

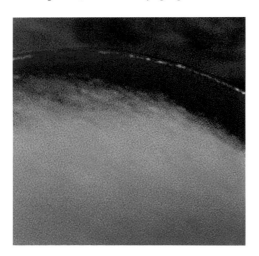

The energy cycle

The Sun is the source of the energy that powers the movements of our atmosphere and produces the temperature ranges that support life. After incoming solar radiation—in other words, energy—penetrates Earth's atmosphere, a complex series of events takes place that together create a stable temperature environment. The solar energy is absorbed and reflected by Earth's surface as well as by atmospheric gases and clouds. Some of this energy makes its way back into space. At Earth's surface, most solar energy is channelled into the evaporation of water, which is subsequently released into the atmosphere when water vapor condenses into liquid water droplets, or deposits back into ice crystals, to form clouds. These clouds reflect and absorb some incoming solar radiation and absorb radiation emitted from the surface. This constant cycling of energy is the genesis of Earth's weather.

WHERE THE SUN'S ENERGY GOES
This diagram illustrates the journey of solar energy as it enters and leaves the atmosphere. Most of the energy radiated back into space—about 64 percent—comes from clouds and gases in the atmosphere. Of the energy reflected from Earth's surface—about 4 percent—an important contribution comes from snow. The balance between incoming and outgoing energy results in Earth's stable temperature. The atmosphere acts as a blanket, warming Earth and maintaining a balance between the amount of solar radiation absorbed and the amount of heat reflected back into space.

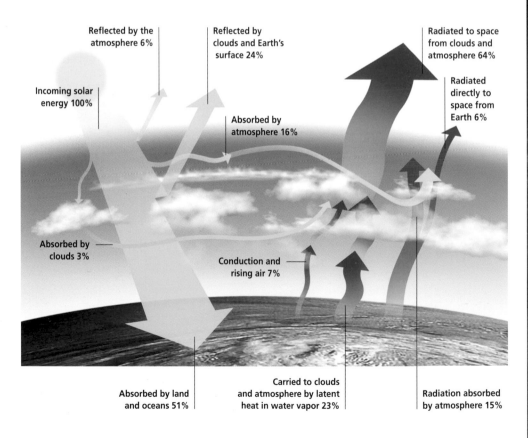

Reflected by the atmosphere 6%

Reflected by clouds and Earth's surface 24%

Radiated to space from clouds and atmosphere 64%

Incoming solar energy 100%

Absorbed by atmosphere 16%

Radiated directly to space from Earth 6%

Absorbed by clouds 3%

Conduction and rising air 7%

Absorbed by land and oceans 51%

Carried to clouds and atmosphere by latent heat in water vapor 23%

Radiation absorbed by atmosphere 15%

← **Visual thermometer** This infrared satellite photograph shows the hottest parts of the Arabian Peninsula in white, the coolest areas in blue. Infrared sensors react to the different heat patterns produced on the surface by solar heating.

↓ **Earth's reflectors** The areas of cloud and snow shown in this satellite view of northern Newfoundland are efficient reflectors of solar energy back into space. Fresh snow reflects 90 percent of the solar radiation falling onto it.

"SEEING" HEAT

Infrared photography is an important tool in monitoring the heat output of Earth's surface. Heat energy is radiated in different amounts from the ground, the oceans and clouds (*see diagram, opposite page*). Even the surface of Earth varies as a radiator: forested regions, for example, have quite different heat-emission characteristics from deserts, snow-covered areas, grasslands or built-up urban environments. Oceans and tropical forests absorb 90 percent of all solar radiation.

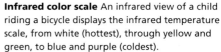

Infrared color scale An infrared view of a child riding a bicycle displays the infrared temperature scale, from white (hottest), through yellow and green, to blue and purple (coldest).

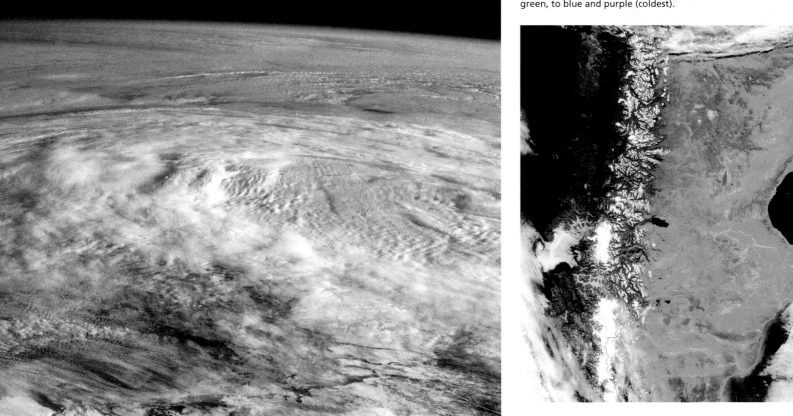

South America revealed Seen from space, the snow-covered Andes Mountains contrast sharply with the lower landscape of Argentina, two regions that have markedly different heat-emission characteristics.

The atmosphere

The atmosphere surrounding Earth contains gases, clouds and other airborne particles called aerosols. Some of the atmosphere's components, such as water vapor and carbon dioxide, also cycle through plants, waterways and oceans. The main gas in the atmosphere is nitrogen, followed by oxygen and other gases, all of which forms a "cocktail" that is unique to Earth. While the atmosphere has no definite ceiling, more than 99.99 percent of its mass lies below an altitude of about 60 miles (100 km). The atmosphere makes Earth habitable by providing protection from the Sun's harmful radiation and helping to maintain warmer temperatures than would occur if it were absent. Despite its importance to life on Earth, our planet's atmosphere is very thin: if Earth were the size of an onion, its atmosphere would be only as thick as the onion's skin. As shown in the diagram at right, this fragile shield is composed of five layers, reaching from Earth's surface to outer space.

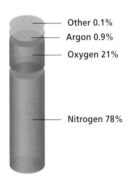

Other 0.1%
Argon 0.9%
Oxygen 21%

Nitrogen 78%

THE AIR WE BREATHE
This graph shows the percentage of gases in the troposphere—where 99 percent of our weather occurs—and the stratosphere. Other gases include carbon dioxide, and traces of neon, helium, krypton, hydrogen and ozone. The proportions of these gases vary at higher levels of the atmosphere. Water vapor comprises up to 4 percent of moist air, which reduces the percentage of the other components. In recent years, increasing levels of carbon dioxide and reduced ozone levels have created concerns about the health of the atmosphere.

↑ **Storm clouds are born**
The flat, anvil-shaped portion of this thunderstorm cloud has formed at the base of the stratosphere, a layer in which temperature increases with altitude.

← **Wonder of the night sky**
Aurora borealis displays, such as this one over northern Canada, occur at an altitude of more than 60 miles (100 km)

⇐ **An atmospheric cross-section** A space shuttle photograph shows sunlight as a blue haze scattered by high-level air molecules. Below is the landmass of northern Africa.

→ **Seeing the invisible** This computer-enhanced image, taken from a meteorological satellite, shows water vapor concentrations in Earth's atmosphere. The dark areas contain less water vapor than light areas and indicate sinking motions in the upper troposphere. Images like this assist meteorologists.

THE MANY-LAYERED ATMOSPHERE

Earth's atmosphere is often described in terms of the vertical variations in air temperature. In the lowest 6 miles (10 km), the air temperature decreases with higher altitude. Since this temperature decrease often enhances vertical atmospheric motions, most of our weather occurs in this zone—the troposphere. Above the troposphere, the temperature increases through the stratosphere. The top of the stratosphere, at an altitude of about 30 miles (50 km), is relatively warm since the Sun's ultraviolet radiation is absorbed by oxygen and ozone. Moving upward from the top of the stratosphere, temperatures decrease through the mesosphere to an altitude of about 50 miles (00 km). Above this, temperatures increase dramatically.

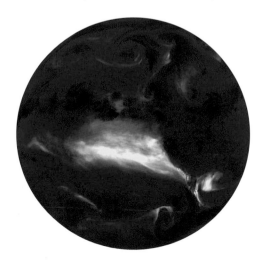

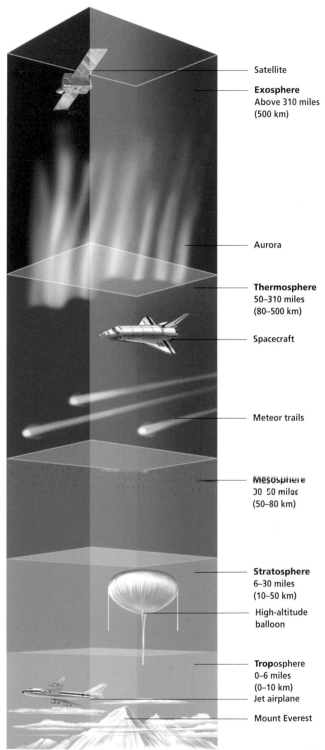

Satellite

Exosphere
Above 310 miles
(500 km)

Aurora

Thermosphere
50–310 miles
(80–500 km)

Spacecraft

Meteor trails

Mesosphere
30–50 miles
(50–80 km)

Stratosphere
6–30 miles
(10–50 km)

High-altitude
balloon

Troposphere
0–6 miles
(0–10 km)

Jet airplane

Mount Everest

Atmospheric pressure

Atmospheric pressure at any point on Earth is caused by the weight of the column of air above that point, as is measured with an instrument called a barometer. When simultaneous pressure readings are taken around the world, and lines of constant pressure—called isobars—are drawn, a fascinating pattern emerges, which can be thought of as the "fingerprint" of the weather. Areas of high and low pressure appear, and these tend to circulate around Earth in well-defined bands. These pressure systems are closely associated with the weather experienced on the ground. High pressure usually produces fine weather, and low pressure is associated with unsettled conditions, sometimes with developing rainfall. As the atmosphere constantly works to restore equilibrium, air moves into low-pressure areas from surrounding areas of higher pressure. This movement of air from high- to lower pressure areas—the flow is always in this direction—is known as wind. As atmospheric pressure is so closely linked with approaching weather, meteorologists observe and record it as a fundamental element in weather forecasting.

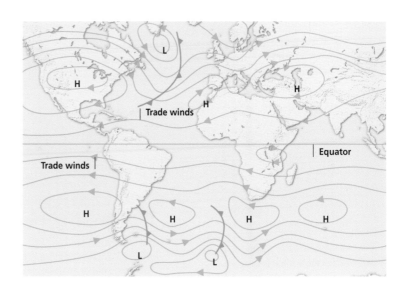

↗ **Where the "highs" and "lows" hover** The areas of high and low pressure that encircle Earth usually form in well-defined bands or pressure belts. In equatorial areas, low pressure predominates, but in midlatitudes, large areas of high pressure lie over both hemispheres. Further belts of low-pressure systems are found toward the polar regions.

↓ **Low pressure, rain likely** These large clouds are vertical formations, capable of producing rainshowers or even thunder. Clouds of this type are often associated with areas of low atmospheric pressure that develops when warm air rises. High pressure results when the air cools and sinks.

HOW DOES A PRESSURE SYSTEM WORK?

This diagram depicts the structure and action of high- and low-pressure cells in the northern hemisphere. Air at the surface is spiraling towards the center of the low in an anticlockwise direction (*right*), before rising and then diverging in a clockwise direction in the upper levels of the atmosphere. The opposite effects govern the movement of the high-pressure system (*left*). Rising air associated with low-pressure systems assists in the production of cloud, and often precipitation. In the southern hemisphere, the rotational directions are reversed, but low-pressure cells are still associated with rising air. Atmospheric pressure is measured in hectopascals (formerly millibars).

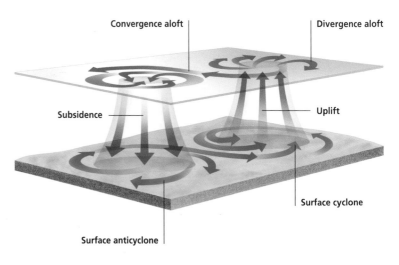

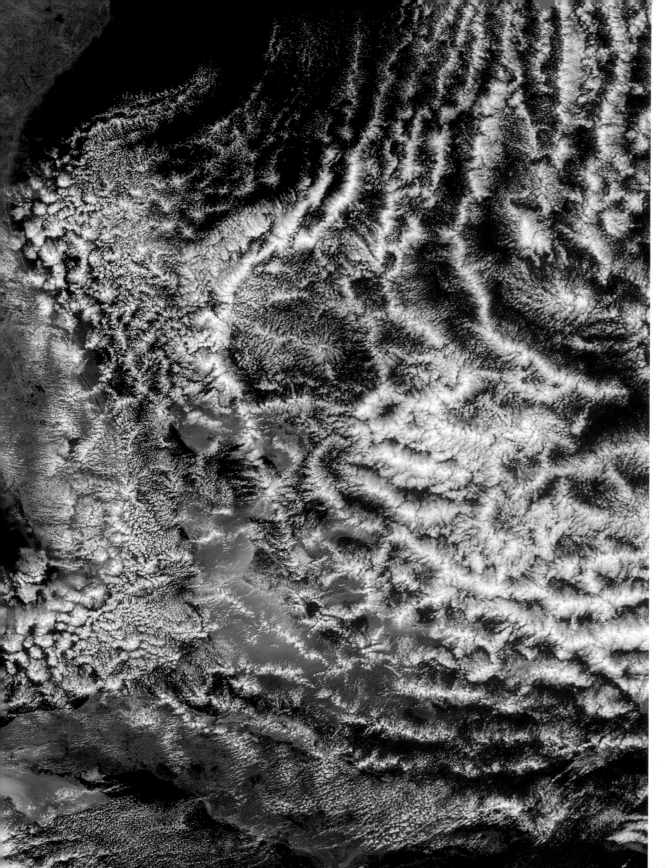

Bahaman clouds
Viewed from above, this cloud pattern is a result of cold air following a low-pressure system being drawn over warmer ocean areas. The speckled cloud formation consists mainly of cumulus clouds. Along with air temperature and humidity, atmospheric pressure is one of the most commonly reported phenomena in media weather reports. While we are conscious of changes in temperature and humidity, we do not readily sense changes in atmospheric pressure. Nevertheless, significant weather shifts result from relatively minor variations In atmospheric pressure.

Global winds

Uneven solar heating of Earth produces varying patterns of airflow, and hence varying weather at different latitudes. Intense heat reaches the tropics throughout the year and produces powerful convection currents. Warm air rises, creating a belt of low pressure around the equator. The air that rises eventually meets the troposphere, where it can rise no farther, gradually cools and sinks back to Earth's surface, at about 30 degrees north and south latitudes. Some of the air from these latitudes, forced out by the sinking air, moves back toward the low pressure at the equator: this airflow is known as the trade winds. The area at the equator where the winds die out is known as the doldrums. The circulations that rise in the tropics, sink at 30 degrees and flow back to the equator are known as Hadley cells. Other circulations continue to move poleward; those that occur between 30 and 60 degrees are called Ferrel cells.

High pressure, blue skies The balmy, tranquil weather associated with tropical islands is often the result of the influence of a high-pressure cell, which is associated with subsiding air and the suppression of cloud development.

Polar cell Cold air at the poles sinks and travels toward the equator before rising upon meeting the Ferrel cell

Jet stream Strong, high-altitude, westerly winds

Hadley cell Warm air rises from the equator and spreads toward the poles before sinking at around 30° latitude north and south. These cells are named in honor of the English scientist, George Hadley, who first described them in 1753.

Doldrums The windless area at the equator

Ferrel cell Some of the air from the Hadley cells continues toward the poles before rising at about 60° north and south. These cells are named after William Ferrel, who first identified them in 1856.

Westerlies Warm, moist winds blow from the west

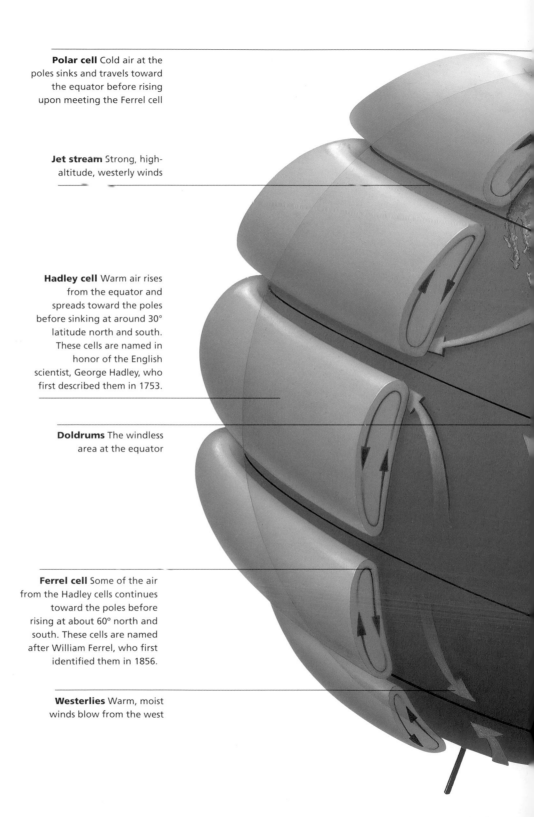

Direction of rotation

Polar easterlies Cold easterly winds blow from the poles to 60°

Northeast trade winds These winds blow toward the equator

WIND AT ITS WORST

In the tropics, low-pressure cells that develop sufficient intensity become hurricanes, often releasing destructive winds and flood rains that can produce widespread devastation. The origin and progress of global airflows are the result of interaction between the Sun and Earth's atmosphere.

THE CORIOLIS EFFECT

The Coriolis effect is best understood by imagining that someone sitting at the center of a moving roundabout (point A in the diagram below) throws a ball to someone sitting at a point on the rim (point B). By the time the ball reaches B, the person on the rim will have moved to position C. To this person, the ball will appear to have curved away from them. Similarly, to us on our spinning planet, freely moving objects appear to follow a curved path. The result, as shown in the diagram below right, is that objects, including weather systems, turn to the right in the northern hemisphere and to the left in the southern hemisphere.

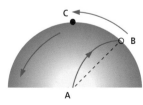

SPINNING PLANET

C

B

A

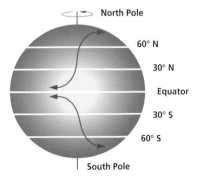

ROTATING WINDS

North Pole

60° N

30° N

Equator

30° S

60° S

South Pole

Jet streams

The wind patterns in the upper reaches of the atmosphere include large, vertical, rotating cells of wind that help redistribute heat away from the equator to higher latitudes. These cells are generally quite stable, but on occasion break down and reform. When this happens they have a major effect on seasonal weather experienced around the world. Periodic droughts and floods have been linked to the behavior of some of these vertical circulations. Fast-moving rivers of air, called jet streams, also race across the high-altitude sky, sometimes reaching speeds of 250 miles per hour (400 km/h). The presence of jet streams was predicted theoretically early in the twentieth century, but not encountered in practice until the air conflict over the Pacific during the Second World War. Monitoring jet streams is an increasingly important element in weather prediction.

UNSEEN BUT INFLUENTIAL

Jet streams consist of predominantly westerly winds, and are caused by strong differences in pressure and temperature in the upper levels of the atmosphere. In winter, when there are greater temperature contrasts, the jets are more pronounced and move toward the equator; in summer they weaken and move poleward. While jet stream winds are very fast, they are relatively narrow: they can be thousands of miles long, hundreds of miles wide, and just a mile or so deep. Long lines of clouds often indicate the presence of a jet stream; the cloud forms when air moves upward and rotates around the jet stream. A knowledge of the location and strength of jet streams is essential to aviation; pilots can reduce flight time by "hitching a ride" on jet streams.

↓ **The jet stream, from above** In 1991, the space shuttle Atlantis captured this image of elongated jet stream clouds stretching across the sky above the Red Sea. The Nile River is at bottom left of the photograph.

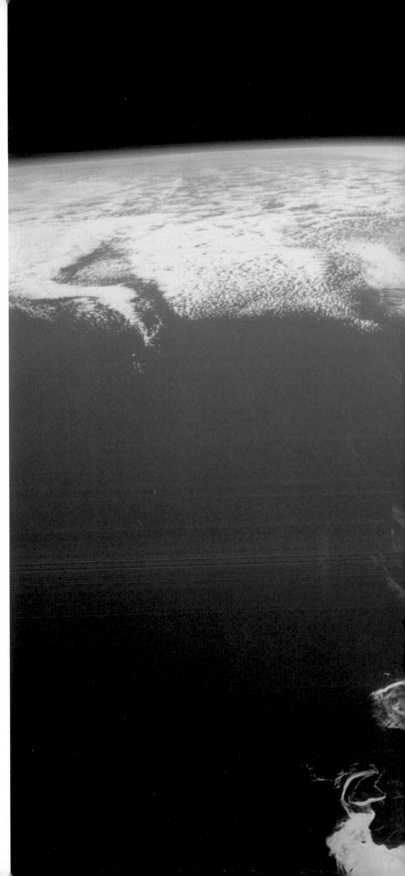

Maritime Provinces, Canada Made visible by high-level cloud, this northern hemisphere jet stream moves over Cape Breton Island in Canada. In each hemisphere, two jet streams snake through the troposphere. While their positions are constantly shifting, jet streams generally coincide with the meeting points of the Hadley and Ferrel cells (the subtropical jet), and the Ferrel and polar cells (the polar front jet). The subtropical jet is the more intense and the higher of the two, about 7 miles (12 km) above Earth. The polar-front jet lies about 3–5 miles (5–8 km) above Earth's surface. Knowing the location and strength of jet streams is valuable to airplane pilots, as planes that can travel in the jet stream save both time and fuel.

Winds at sea

The global wind patterns that circulate around the planet have played a major role in maritime exploration and trade over the centuries. Sailing-ship captains usually made good progress in the trade-wind belt, as they knew that helpful winds could be expected. The doldrums, on the other hand, were dreaded because they could result in ships being becalmed for long periods, with the crew members—usually from cool-climate countries—having to endure uncomfortably humid tropical conditions. In the days before weather-forecasting services, many ships were lost at sea after encountering unexpected gales. The "roaring forties" were particularly feared, because of the frequent savage westerly winds that sweep across these latitudes. The waters near both the poles also had a bad reputation for extreme weather. The global circulation system is critical to the weather engine. From the stifling doldrums, the warm mass of equatorial air may travel to the poles, to the edge of space, over thunderclouds. At times, it creates a gentle breeze; at others it whips up a hurricane. For sea goers, the power of wind can never be underestimated.

The Nantucket rides it out A painting by Edwin Walter Dickinson depicts the unfortunate steamer *Nantucket* enduring gale-force conditions. Even vessels with powerful modern engines can find themselves at the mercy of heavy weather.

THE "ROARING" SOUTHERN LATITUDES

In the region between latitudes 40° and 60° south, westerly winds are frequent, often reaching gale force. At latitude 60°, there is no major landmass to interrupt the wind as it blows around the planet and the result is huge ocean swells that travel constantly from west to east. The distance between crests can be over three quarters of a mile (1.2 km) with a wave height of up to 70 feet (21 m). When these swells are not breaking, ships can travel up and over them in safety.

↓ **Into the breaking swell** The icebreaker *Kapitan Khlebnikov* plows through wild seas off South Georgia Island in the South Atlantic Ocean. Gale-force conditions and rough seas are common in these latitudes, particularly during winter.

↓ **How Earth's winds blow to a pattern** The light equatorial winds are surrounded to the north and south by the trade winds. This zone of light equatorial winds migrates north and south with the seasons, triggering the onset of monsoons.

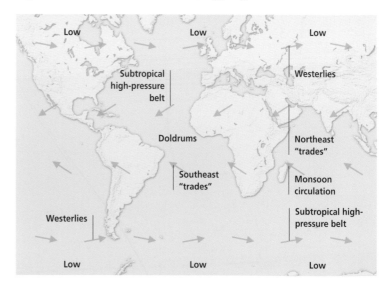

IN THE DOLDRUMS

If there was anything that sailing-ship crews dreaded as much as gales, it was the opposite situation—becoming trapped in a region of calm conditions, when their ship could not make any headway. Such conditions were often encountered around the tropics in the light wind belt called the doldrums. In his famous epic "The Rime of the Ancient Mariner," Samuel Coleridge describes being becalmed:

Day after day,
*   day after day,*
We stuck, nor breath
*   nor motion*
As idle as a
*   painted ship*
Upon a painted ocean.

A long period in the doldrums resulted in serious problems—food and water shortages, and perhaps even more important, a drop in crew morale.

Nary a breath of wind This becalmed yacht drifts listlessly in windless, calm conditions, a situation not uncommon as vessels encounter the low-pressure belt around the tropics.

Frontal systems

Weather monitoring and prediction in the midlatitudes is usually related to the movement of frontal systems. There are three main types of fronts—warm fronts, cold fronts and occluded fronts. A warm front shows the boundary of an approaching warm air mass and is represented on weather maps as a line with filled semicircles on it. Similarly, a cold front shows the boundary of an approaching cold air mass and is drawn as a line with filled triangular barbs on it. An occluded front is a frontal boundary that has been lifted off the surface. It is shown as a cross between a warm and cold front. The stronger the various fronts are, the more severe the weather tends to be, with a greater temperature change across cold and warm fronts. Fronts rarely occur in the tropics where the temperature differences are minor.

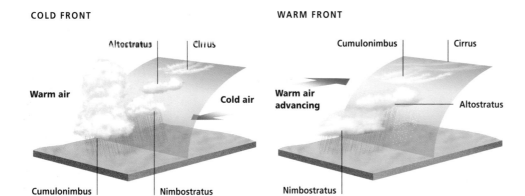

COLD FRONT

Altostratus · Cirrus

Warm air · Cold air

Cumulonimbus · Nimbostratus

WARM FRONT

Cumulonimbus · Cirrus

Warm air advancing · Altostratus

Nimbostratus

COLD AND WARM FRONTS

In a cold front (*above, far left*), dense cold air pushes into a warmer air mass, causing strong lift near the surface just ahead of the cold air. This lifting may produce showers and thunderstorms. These showers decrease as the cold air deepens behind the front. In a warm front (*above left*), warm air slides up and over a denser, pre-existing, cooler air mass. This tends to produce a more gradual thickening and lowering of cloud as the surface warm front approaches.

↑ **A clue in the clouds** Developing "turrets" in these cumuliform clouds indicate that the atmosphere is becoming more unstable; this promotes further vertical growth of the clouds.

← **Squall coming** A sudden lowering of cloud, and the appearance of smooth, rotating cloud features, indicate that a swift, possibly violent wind change is about to arrive. A sudden rise in atmospheric pressure and a drop in air temperature commonly accompany the passage of squall lines like this one. Thunderstorm activity is also likely with such changes, generating strong and gusty winds.

OCCLUDED FRONTS

A frontal occlusion is a phenomenon that may take up to 48 hours to run its course. The first stage of the occlusion process sees a warm front well ahead of an approaching cold front. Initially, there are three distinct air masses with cooler air ahead of the warm air mass that arrives with a warm front. Next comes the cold air behind the cold front. The slower movement of the warm front allows the cold front to catch up. After one or two days the cold front overtakes the warm front. This forces the warm air off the surface and so the surface effects of the warm front vanish. Some rain may still occur from the front that is now called "occluded," although this is usually not heavy.

This sequence of events is more common in the northern hemisphere than the southern hemisphere because well-developed warm fronts are less common in the south.

OCCLUDED FRONT

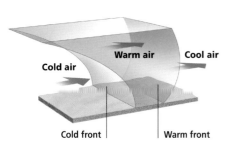

Cold air   Warm air   Cool air

Cold front   Warm front

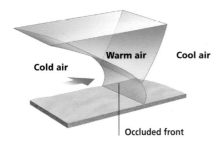

Cold air   Warm air   Cool air

Occluded front

→ **Quick change** This space shuttle view of a squall line, or line of thunderstorms, over the Atlantic Ocean southeast of Bermuda illustrates how quickly the weather can change from fine to stormy.

Highs and lows

The high-pressure belts that circle the globe in the midlatitudes play an important role in determining the weather. When high pressure lies overhead, conditions tend to be fairly quiescent with light and variable winds. There can be fog and low cloud if there is plenty of low-level moisture, although ridges can also produce brilliant, sunny skies as the air dries out. The belts of high pressure are periodically broken by sudden incursions of cold, polar air that can produce cold, wet and windy weather. At other times, particularly during the warmer months, incursions of tropical air can generate bouts of thundery weather. The transition from fine to stormy weather can sometimes be very rapid, with some areas more susceptible than others to changeable weather. Locations in proximity to mountain ranges often experience variable conditions over a relatively short time. Mountains cause air to rise, enhancing cloud formation and precipitation. Most of the precipitation will fall on the windward side.

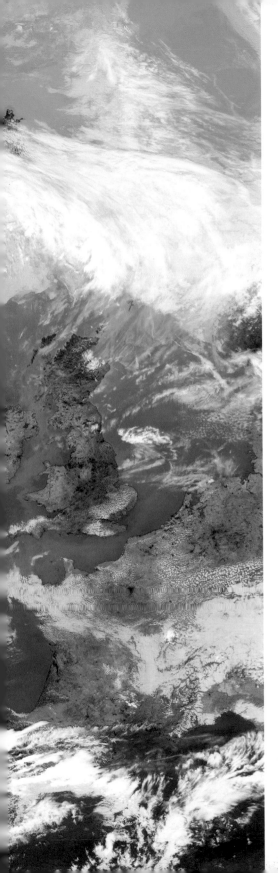

LOW-PRESSURE SYSTEM

Low-pressure systems form when two air masses at different temperatures interact. In the diagram at right, a cold air mass meets a warm air mass (1). Gradually, the warm air rises above the cold air, creating an area of low pressure into which the cold front moves. The rising warm air creates clouds and rain, and the fronts begin to rotate (2). Over time, the faster-moving cold front catches up with the warm front. As pressure decreases under the rising air, precipitation increases (3). When the cold front catches up with the warm front, an occluded front forms. Windy, unsettled weather results (4). When the occluded front is fully formed, it cuts off the supply of warm air, and winds and precipitation subside. If the two air masses reform, the cycle begins again (5). This process is known as cyclogenesis.

Areas of low and high pressure form and break up in response to a complex set of events, including the Sun's heating of the atmosphere, Earth's rotation, and the interaction of the atmosphere with landmasses and the oceans. As they play such a large part in determining the weather, identifying and monitoring pressure systems is fundamental to accurate weather forecasting.

→ **The pressure cycle** Lows are areas where buoyant air is rising, and are associated with cloudy or rainy weather. Highs are areas where dense air is sinking, and are associated with fine weather. Although the concept of high- and low-pressure zones has been known since the seventeenth century, full understanding of the complexities of this global system is relatively recent.

← **Capturing the moment** A dramatic satellite image shows frontal cloud spiraling into the center of an intense low-pressure system over the North Atlantic; Europe has the relatively clear skies of a high-pressure system.

←← **Whirlpool of clouds** The graceful swirling cloud spirals of an intense low-pressure system high over the North Pacific Ocean are evident in this space shuttle photograph.

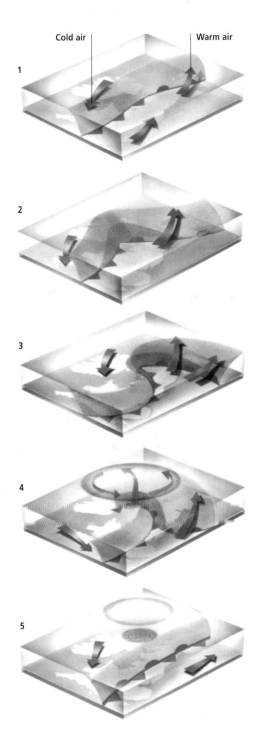

Cold air   Warm air

1

2

3

4

5

Heating and cooling

The diurnal heating and cooling cycle of Earth's surface has a marked effect on the weather. As the surface heats up, thermals form and the air just above the ground becomes more turbulent. This brings stronger winds from the lower atmosphere down to the surface, with winds likely to become gusty at times.

Sea breezes are a feature of coastal locations. Overnight, Earth's surface rapidly loses heat, particularly if the skies are clear. Temperature inversions can form, with winds below the inversion becoming light and variable. Soon after sunrise, the heating of the ground begins the cycle again, thermals generate, inversions dissipate, and upper-level winds drop to the surface.

THE DAY-AND-NIGHT DANCE OF THE WINDS

In hilly terrain, upslope (anabatic) and downslope (katabatic) winds are common. The rising Sun heats hills more rapidly than valleys. This causes thermals to form over the hilltops. Air from the valleys flows up the slope to replace the rising air in the thermals. Overnight, this cool air flows down the hillsides into the valley.

↑ **Alone in flight** An albatross glides over a beach on Midway Island in the Pacific Ocean. Thermals assist seabirds in their long-distance migratory routes.

← **Catching the wind** For ocean-racing yachts that compete in round-the-world events, global winds are of the utmost importance. Some teams employ meteorologists to assist in their route planning.

↓ **Night-time cooling** Photographed by an Earth-orbiting satellite, the night-time clouds over southeastern Alaska, USA, align themselves with the cold, downslope (katabatic) drainage flows off the rapidly cooling slopes of the mountains.

LAND AND SEA BREEZES

Coastal locations commonly experience a reversal of winds during the day and again overnight. Sea breezes form when the land heats up more quickly than the ocean, causing pressures to fall over the land and the air to rise. Cooler air rushes in from the water to replace the rising, warmer air, creating a sea breeze. Sea breezes normally reach peak strength in the late afternoon. Overnight, the land cools more rapidly than the ocean, so its surface air, cooler than the ocean air, drains off the land and onto the sea.

↑ **Riding the breeze** Windsurfers enjoy the gusty sea breeze offshore from Noumea in New Caledonia. Soaring behind them is the Tjibaou Cultural Center, designed by Renzo Piano and named in honor of the former leader of the Kanak people.

AFTERNOON

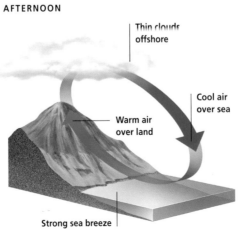

Thin clouds offshore

Cool air over sea

Warm air over land

Strong sea breeze

NIGHT

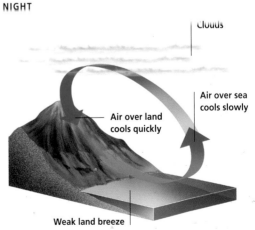

Clouds

Air over sea cools slowly

Air over land cools quickly

Weak land breeze

The monsoon

Monsoons, from the Arabic word *mausim* meaning "season," are the seasonal changes in wind, often rain bearing, that are experienced in many tropical regions. Originally used by seafarers crossing the Arabian Sea to describe the six-monthly reversal of the winds from the northeast to the southwest, the term now applies to the two main seasons in the tropics. Although monsoons affect the continents of Asia, Africa and Australia, nowhere are the monsoonal winds and rainfalls so dominant as over southern and southeastern Asia. In India more than 75 percent of the annual rains fall during the southwest monsoon. Cherrapunji, in northeastern India, provides a good example of this. The town's average rainfall for the month of December is only ½ inch (13 mm) compared with the June average of 106 inches (2695 mm) when the monsoon is normally at its height. The resulting annual rainfall totals are among the highest in the world, exceeding 500 inches (12,700 mm) in some locations. Half the world's population relies on monsoon rains for vital water supplies. Indeed, India, Bangladesh and Pakistan are together described as the monsoonal subcontinent.

WINTER MONSOON

SUMMER MONSOON

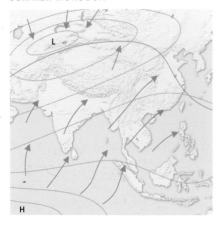

THE BIRTH OF A MONSOON

The monsoons of southern and southeast Asia are driven by seasonal changes in the global weather patterns. During the northern hemisphere winter, the Siberian high-pressure system intensifies, to become the world's strongest. This produces strong northeasterly winds that blow toward the equator. These winds remain dry until they pick up moisture over the South China Sea and become the northeast monsoon of southeast Asia. The southwest monsoon of the summer months forms when an intense heat low forms over central Asia. This draws in the southeast trade winds that become southwesterly after they cross the equator. These winds can bring heavy rains to south Asia. The convergent zone between the trade winds and the monsoon is called the Intertropical Convergence Zone. The rains produced by monsoons are essential for the survival of billions of people. With the arrival of the monsoonal rains, fields that have been barren for months become fertile acres. Although the monsoon winds are an annual occurrence, there are times when the rains are late, or lighter or more sporadic than normal. Crops fail and millions face starvation.

→ **Seasonal hazard** Cyclists and tricycle rickshaws in Bangladesh battle along flooded roadways during a monsoon. Evacuation of residents and disruption to commerce and schooling are common at the height of the monsoon season. Despite the short-term problems, however, the monsoon is a life-giving phenomenon, replenishing water supplies for the dry season ahead. From southwest Arabia to the Indian subcontinent and southeast China, crops rely on monsoonal rainfall.

→ **Forecast: rain** Modern techniques allow monsoonal downpours like this one at Simla in India to be predicted more accurately than in the past. On the Indian subcontinent. monsoonal rains are at their height from June to October, when winds turn again to the northeast, humidity falls and rain eases.

← **Celebrating the monsoon** Villagers walk through rain in Pandwa village, Dang district, south of Ahmadabad, India. The monsoon has a significant role in the lives of the tribal people across the western Indian state of Gujarats Dang, where people depend on forest produce. The woven headgear—both traditional and decorative—provides protection from the elements.

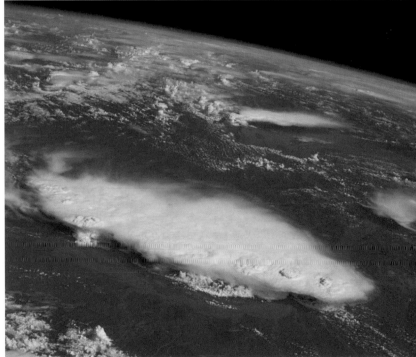

THUNDER FROM ABOVE
Severe thunderstorms are common across the Indian subcontinent during the monsoon season. Here, a group of well-developed thunderstorms can be seen at the center of the image, with some less intense, decaying cells at the top. Before the days of satellite and radar, people used natural signs to predict the start of the monsoon season. Today's meteorologists have sophisticated equipment with which they can predict the monsoon's arrival to within a few days.

Local winds

The combination of local temperature contrasts and the shape and size of physical barriers, such as mountain ranges, escarpments and valleys, produces a wide range of fascinating, but sometimes devastating, local winds. Because of their dramatic effects, often repeated several times a year, many of these winds have been given a characteristic name. In France and the Cote D'Azur, there is the mistral that has been known to reach speeds greater than 80 knots. The USA and Canada both experience the chinook—a warm and gusty wind that roars down the eastern slopes of the Rocky Mountains. And California has its hot, often dust-laden santa ana that brings the desert to the coast, triggering numerous wildfires in the process.

FAMOUS LOCAL WINDS

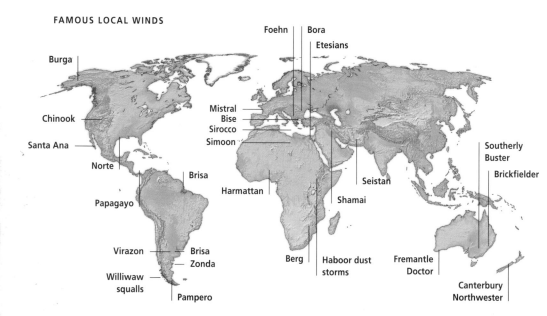

← **Dawn in the Mojave Desert** The steep ranges of the Sierra Nevada, California, USA, generate strong wavelike airflow whenever the midlevel winds are strong and blow at right angles to the ranges. These mountain waves can generate turbulent conditions for aircraft .

→ **Fanned by the mistral** A fireman attends a fire fanned by a strong mistral wind in Saint Chamas, France. Fires of this kind can be destructive to life and property alike.

→ **Death Valley desert** Wind whips sand over the dunes in Death Valley National Park, California, USA. Long periods of extreme heat and little rain (on average, less than 2 inches [5 cm] a year) help to produce the fine sands of desert dunes. These sands can be carried great distances when strong winds blow across the region. Despite the harsh environment, more than 900 species of plants grow within the park.

← **Santa ana dust storm** The dry and dust-laden santa ana winds of southern California, USA, provide all the ingredients necessary to turn minor fires into destructive infernos within the space of a few hours.

Extreme winds

Local winds can be either extremely hot or cold, depending upon their origin and method of formation. When the air source is very cold, such as near the Arctic or Antarctic regions, the winds can be freezing cold, even when experienced a thousand miles away from their origin. In stark contrast, strong winds blowing across sandy midlatitude deserts can carry dust for thousands of miles, bringing hot and dusty weather to normally moist, temperate regions. Huge dust plumes from the Sahara have been identified over the eastern Atlantic Ocean, and dust from central Asia reaches the northwestern Pacific. Thunderstorm winds cause walls of dust to travel across the desert; the Sahara alone produces 300 million tons of dust a year.

THE PAMPERO AND BORA

The passage of strong frontal systems over Argentina periodically produces the pampero—a very cold, southwesterly wind that blows over the pampas plains. The Andes Mountains help to funnel cold, dry Antarctic air northward across Argentina. Severe thunderstorms with strong squalls sometimes herald the arrival of the pampero. A severe form of the pampero—the pampero sucio—produces a dust storm. The Adriatic region of Europe has its own cold wind, known as the bora. The source of the air from over Russia is so cold that the wind does not warm up significantly as it descends to the sea.

SOUTHERN GALES

Strong winds are characteristic of the Antarctic climate and are caused by cold inland air flowing, under the influence of gravity, down the ice sheet to the coast. Howling blizzards with wind speeds of 100 miles an hour (160 km/h) are frequent, usually accompanied by heavy, swirling snow that reduces visibility. The gales of the Southern Ocean, where strong low-pressure centers develop, have long been the bane of mariners. Even today, with every technological advantage, ships are vulnerable here.

↑ **Desert dust** When windstorms rage across the Sahara Desert, dust can be carried thousands of feet up into the atmosphere and swept thousands of miles off the coast. Fine dust is ultimately deposited into the oceans, where it becomes an important nutrient source for marine organisms.

← **Polar survivors** Assisted by the right weather pattern, extremely cold air draining off the elevated Antarctic Plateau inland from the Weddell Sea produces some of the world's strongest blizzards. Only the hardiest of animals, such as these emperor penguins, can survive in these conditions. This is the most extreme form of katabatic wind.

→ **Near Timbuktu** This Tuareg camp outside Timbuktu in Mali has been hit by a severe dust storm blown up by harmattan winds. These parching winds are a continental expression of the trade winds that encircle the globe. They dominate the Sahara Desert and, particularly from December to February, bring extremely dry, warm air to much of North and West Africa, often providing welcome relief from high humidity. Dust transported by the harmattan can reach South America, far from its source.

→ **Antarctic blizzard** The steep slopes around the edge of the Antarctic continent accelerate cold katabatic winds down toward the warmer waters of the Southern Ocean, making coastal blizzards a frequent event in places like Butson Point, a northeast glacier of Antarctica. Long, savage and almost perpetually dark winters combine with these hurricane-force winds to make it dangerous to venture outside.

Ocean currents

Global ocean currents have a huge impact on the weather experienced over nearby lands. The air above cold currents carries little moisture, which helps to create deserts on the west coasts of continents at midlatitudes around the world. The contrast between cold ocean temperatures and the warm land also produces regular sea breezes along these coasts. Warm currents transfer abundant moisture to the winds that blow over them, feeding the weather systems that can bring rain and thunderstorms to eastern coasts.

PERPETUAL MOTION: THE GREAT OCEAN CURRENTS

The major ocean currents greatly affect the weather and climate of the regions they flow past. Warm currents, such as the Gulf Stream and Californian Current, can keep temperatures in near-polar regions much warmer than they would otherwise be, and are great moisture sources. Cold currents, such as the Peru and Benguela currents, tend to reduce rainfall over nearby land.

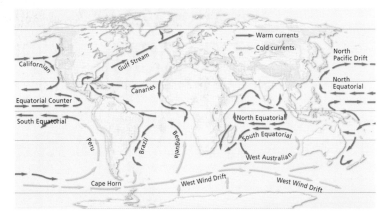

← **Gulf Stream over Florida** The Gulf Stream is the strongest ocean current in the northern hemisphere. It carries about 30 billion gallons (135 billion liters) of water every second.

→ **Riding the current** Some sea turtles migrate long distances—up to 2800 miles (4500 km)—from their nesting beaches to feeding grounds. How they navigate is not fully understood, but it seems they rely on currents to guide them.

→ **Atacama Desert** Very cold waters of the Peru Current off Mejilones Peninsula, Chile, produce the extremely arid conditions of the world's driest desert, the Atacama. Fog and low stratus cloud often form over these cold waters, as is evident in this satellite image.

←← **Ocean meets desert** The Atlantic Ocean meets the Namib Desert on the western coast of Africa. Temperatures along this coast are mild to cool because of upwellings of cold water from the Benguela Current. This cooling brings high humidity and as many as 250 days of fog each year. Desert dwellers and cold-water animals such as seals share this unique environment.

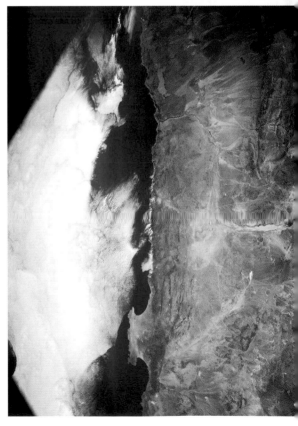

Previous page Water is an essential element
for life on Earth, and exists in many forms.

Weather in action

The beauty of a cloud formation, the symmetry of a rainbow, the life-giving power of rain and the eerie quiet of a winter snowstorm—all these are expressions of weather in action. And all owe their origin to water in the atmosphere.

The nature of water

Water, one of the most common elements on the planet, is a relatively simple but flexible substance. One molecule of water consists of two atoms of hydrogen and one atom of oxygen. This very basic molecular structure gives water some unique chemical and physical properties that provide it with an extremely important role in helping maintain Earth as a place of habitable environments. The basic elements of hydrogen and oxygen come together in such a way that they produce mutually attractive forces between water's molecules, known as hydrogen bonds. The most obvious property to result from this bonding is the flexibility that allows water to exist in three quite different forms—solid, liquid or gas—depending on the temperature and the air pressure in which it is exposed. Within these three forms, water manifests itself in a remarkable variety of ways.

PORTRAIT OF A MOLECULE
A water molecule consists of two hydrogen atoms and one oxygen atom. The side of the molecule with the hydrogen atoms has a positive electric charge, while the other side is negatively charged. This difference provides an electrical attraction between molecules, known as hydrogen bonding.

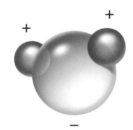

A FLEXIBLE ELEMENT
Earth is the only known planetary body with the comparatively narrow range of temperatures that allows water to exist in solid, liquid and gaseous forms within the general environment. The speed of movement of the water molecules tends to accelerate as an increase in the atmospheric temperature sets in train the gradual transformation of this versatile element from solid (ice) to liquid (water) and finally to gas (water vapor).

Solid Water molecules in the rigid lattice known as ice are held together in a hexagonal arrangement that produces the familiar pattern of six-sided ice crystals.

Liquid At normal temperatures, there are fewer hydrogen bonds between water molecule clusters and the molecules may move more freely as fluid.

Gas High temperatures allow water molecules to become agitated, break most of their hydrogen bonds and move freely as water vapor. Small amounts occur in the atmosphere.

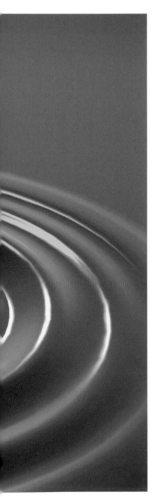

← **Drifting reservoirs** When water evaporates from Earth's surface and condenses in the atmosphere, it forms clouds—an aggregation of minute droplets or ice crystals suspended in the atmosphere at varying altitudes. These common stratocumulus formations are not rain-bearing clouds.

← **From liquid to solid** Dripping meltwater from snow or ice turns solid again in the form of an icicle when the surrounding air temperature starts to approach the freezing point of water. The icicle gradually becomes longer and wider—but mainly longer—as the water continues to flow down along its surface and eventually freezes in new layers that extend along the length of the icicle.

↓ **On the boil** Bubbles of vapor move up rapidly through boiling water. Since the boiling point depends upon the air pressure above the water surface, lower boiling points occur at high altitudes.

← **Drip, drip, drip . . .** The spherical shape of droplets is the result of water's surface molecules being attracted to each other with greater force than those beneath, forming what is known as surface tension. The spherical shape represents the smallest possible surface areea for a given size.

The water cycle

Earth is often called the water planet because of the relatively large amounts of water that exist in three physical forms—liquid, solid (ice) and gas (or vapor)—in a variety of reservoirs. Water can also change its form as it flows constantly between these reservoirs in an elaborate circulation regime called the water cycle. The largest reservoir, containing approximately 90 percent of all water on Earth, is known as the the hydrosphere; this includes the oceans and other large water bodies such as lakes and rivers. The second-largest reservoir is the cryosphere; this is the solid water that makes up the polar ice caps, glaciers and permanent snowfields. Water can also reside in the lithosphere, the upper part of Earth's crust, as groundwater. Small amounts of water are also found in the atmosphere, either as water vapor or as liquid droplets and ice crystals that form clouds.

The Sun helps power the never-ending water cycle, where water flows from one reservoir to another, often undergoing a dramatic physical change. As the water cycle is a closed system, the total amount of water on the planet is relatively fixed. Within the system, water is constantly recycled as evaporation, condensation, precipitation and runoff cause it to flow between the various reservoirs.

THE IMPORTANCE OF OCEANS

Oceans cover 71 percent of Earth and contain 97 percent of its water. They are intimately connected with the weather. Oceans heat and cool more slowly than landmasses, and moderate temperature changes on land. Turbulence and currents within the oceans distribute heat changes through a vast body of water. In addition, the great ocean conveyor belt endlessly circles the planet, transporting warm and cold water around the globe.

↓ **Captive sea** The Caspian Sea evaporation basin in northwest Asia as seen from space. The sea loses its water by evaporation and this maintains its slight salinity. Close observation of such large water bodies provides early evidence of changes in the delicate balance of the water cycle.

→ **Rivers of ice** Snow falling onto glaciers compacts into layers that form ice which, under pressure, flows toward the ocean where it melts or calves. About 77 percent of Earth's freshwater is held in ice sheets and glaciers, and depositing ice and water into the oceans is part of the global water cycle.

SEAWATER AND FRESHWATER

Water, water everywhere, but *hardly* a drop to drink. There is a vast amount of salt water on Earth that is not usable for human consumption, while only small amounts of freshwater exist in a usable form. The diagram below illustrates how Earth's tiny percentage of freshwater exists mainly as ice. Smaller amounts are found underground and in freshwater lakes and rivers.

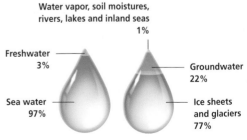

Water vapor, soil moistures, rivers, lakes and inland seas
1%

Freshwater
3%

Groundwater
22%

Sea water
97%

Ice sheets and glaciers
77%

THE WATER CYCLE

Water evaporates from the various liquid reservoirs on Earth's surface—oceans, rivers and lakes—into the atmosphere where it condenses into clouds. Eventually the water leaves these clouds in a process known as precipitation, returning to the surface as rain and snow. More water evaporates from the oceans than is replenished over them by precipitation, but more precipitation falls on the world's landmasses than is lost by evaporation from the land. This endless process maintains a balance: the excess water that falls on the land either runs off in rivers and streams that eventually flow back to the oceans, or it percolates into the ground where it takes a slow, subterranean route toward the oceans. Once there, the cycle of evaporation and precipitation begins again, and the cycle continues. A balanced water cycle is critical to the health of planet Earth.

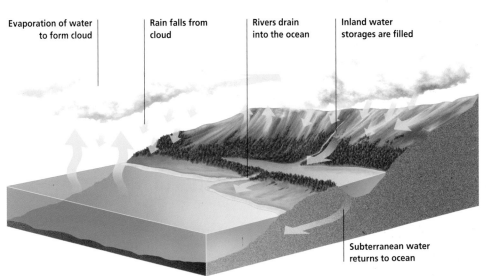

Evaporation of water to form cloud

Rain falls from cloud

Rivers drain into the ocean

Inland water storages are filled

Subterranean water returns to ocean

Humidity

Humidity refers to the amount of water vapor—a colorless gas—in the atmosphere. Vapor levels depend upon temperature and vary considerably across the planet, from barely discernible amounts in polar latitudes to nearly 4 percent of the air by volume in tropical regions. As the air temperature increases, more water molecules evaporate, some of them eventually condensing into liquid. When an equilibrium occurs between the evaporation rate and the condensation rate, the air has reached saturation point and can hold no more vapor.

→ **In the jungle** The near-saturated air of this humid rain forest has been forced upward and cooled by the mountain barrier to form light, mistlike clouds. Humid rain forests provide ideal conditions for plant and animal life.

↓ **Dawn in the Amazon** The Amazon Basin has a pure equatorial climate, with high rainfall and no defined dry season—perfect conditions for high humidity.

Morning mist When night air cools and becomes saturated with water vapor, the vapor may condense to form fog or mist that remains until the Sun warms the air and the vapor evaporates.

MEASURING HUMIDITY

Air is saturated—containing a high level of water vapor—when a dynamic equilibrium is attained between the rates of evaporation and condensation. The air temperature at which saturation occurs is called the dewpoint. In this graph, if an air mass holds half a cubic inch of water vapor per cubic yard of air, the dewpoint will be 52.5°F (11.4°C).

Benign vortex An unusual, tubular cloud, where condensed water vapor has formed a narrow column, shows that condensation can sometimes be a highly localized phenomenon. Not all vortices become destructive tornadoes.

HUMIDITY MADE VISIBLE

A specialized sensor onboard an orbiting satellite provides a snapshot of the world's humidity at any given moment. The dark blue area around the equatorial belt depicts the highest concentrations of atmospheric water vapor, while light-colored areas in the polar regions indicate relatively little vapor.

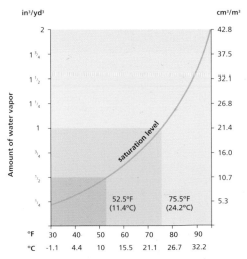

in³/yd³   cm³/m³

Amount of water vapor

saturation level

2		42.8
1 ³⁄₄		37.5
1 ¹⁄₂		32.1
1 ¹⁄₄		26.8
1		21.4
³⁄₄		16.0
¹⁄₂	52.5°F (11.4°C)	10.7
¹⁄₄	75.5°F (24.2°C)	5.3

°F	30	40	50	60	70	80	90
°C	-1.1	4.4	10	15.5	21.1	26.7	32.2

Air temperature

Dew and frost

The dew or frost that forms on surfaces such as lawns and automobile windshields is the result of air near Earth's surface becoming saturated with water vapor. Dew and frost manifest themselves during the night when moist air near the ground cools to saturation at the dewpoint—the temperature at which water vapor begins to condense. Dew develops when water vapor condenses at dewpoint temperatures above the freezing point of water (32°F or 0°C), while frost occurs when saturation occurs at temperatures below the freezing point. Hoarfrost is usually white and feathery, while a denser form of frost, called rime, occurs when supercooled water droplets from fog or clouds freeze on a surface.

← **Pearly dew** Dewdrops are captured in the net of a cobweb when the temperature falls and the moisture in the air condenses. These drops are spherical as a result of surface tension.

→ **The icy beauty of frost** When supercooled water droplets from fog or low clouds freeze on trees and other plants they form a heavy white frost covering known as rime.

← **Cool night, morning delight** Delicate dewdrops on a leaf stem grow bigger each time condensed water runs along the stem. Evapotranspiration from plants adds water vapor to the atmosphere, contributing to dew formation by helping saturate the air just above the ground. The conditions needed for the formation of dew are similar to those for fog, and the two often occur together.

AN OVERNIGHT TRANSFORMATION

The conditions for the formation of dew and frost can be quite similar. Both phenomena generally require a clear night with little or no wind, and an atmosphere that contains sufficient moisture to allow condensation to occur.

In the case of dew, if the ambient air temperature remains above 32°F (0°C) then liquid water droplets will form on exposed surfaces. When the air temperature drops below 32°F (0°C), water vapor in the atmosphere can immediately form ice crystals without first condensing as a liquid. These crystals, when magnified, reveal exquisite jewel-like patterns that branch outward from the grass stems and edges of leaves where they have formed.

Although dew is often associated with cold climates, it also occurs in hot and humid regions. Overnight dew formation is a life-giving source of water for many desert-dwelling plants and animals. Dew is most common in coastal areas.

↑ **Dewdrops** When the air temperature cools to a dewpoint that is above 32°F (0°C) individual water molecules in water vapor condense into tiny droplets. These clump together on exposed surfaces to form larger drops called dew.

↑ **Frost** When the air temperature cools to a dewpoint that is below 32°F (0°C) water vapor changes to ice crystals on exposed surfaces in a process called deposition.

← **Ephemeral crystals** Feathery translucent ice crystals, often called hoarfrost, form on objects such as leaves when the air is cooled overnight to saturation at temperatures below the typical freezing point. Such frosts thaw into water as the Sun's rays warm the atmopshere. Despite its beauty, frost poses a hazard to motorists when roads become slippery, and can destroy fruit and vegetable crops when buds are damaged.

Cloud formation

A cloud is a collection of water droplets or ice crystals that have become dense enough to be visible. The formation of these droplets or crystals usually requires the presence of tiny airborne nuclei in the atmosphere to serve as sites for the condensation or deposition to occur. Clouds usually form when moist air is cooled to saturation, followed either by condensation to form water droplets or by deposition to produce ice crystals. Temperature normally decreases with altitude. Therefore, clouds forming at high levels of the troposphere tend to be ice-crystal clouds, while those at lower levels are more typically composed of water droplets.

The rising motion of air to produce a cloud can be associated with convection or mechanical or dynamic uplift. Convection occurs when warm air becomes more buoyant than its surroundings and moves upward. Mechanical, or orographic, uplift happens when air moves over mountain barriers. Dynamic uplift is associated with large-scale air movement into surface low-pressure systems or along frontal surfaces where the air density is uneven. In brief, the basic processes for cloud formation are: lifting, cooling and condensation.

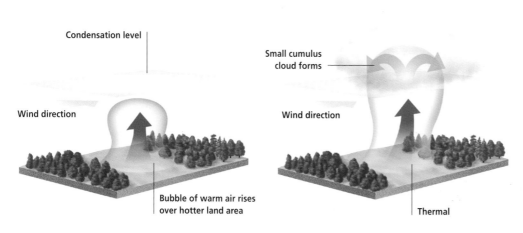

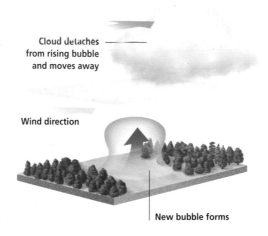

HOW CLOUDS FORM

Clouds form when the convective condensation level reacts with rising air. When air over the warmest part of the land rises through the condensation level, it becomes saturated and condenses to form cloud droplets. An air mass will continue to rise as long as its temperature is higher than that of the air around it. If this situation persists as the air mass rises, conditions are said to be unstable. But if an air mass quickly reaches the temperature of the surrounding air, and stops rising, conditions are said to be stable. A rising air mass cools at a rate of 5.4° F for every 1000 feet (9.8° C per km). Therefore, if we know the temperature of a rising air mass at ground level, and the temperature of the air at different levels of the troposphere, we can calculate how far the air will rise.

↑ **Variations of form** Anvil tops of cumulonimbus clouds float over cumulus congestus clouds. These clouds near the top of the troposphere contain ice crystals and differ from the puffy, lower-altitude cumulus congestus clouds that contain liquid water droplets.

↑ **Hidden valley** Cool air draining into this valley has formed a mixed blanket of fog and cloud. Distinguishing between cloud and fog can be difficult in mountainous regions where clouds obscure the mountains because of the effects of orographic lifting.

→ **Daytime fluffy clouds** These fluffy, puffy clouds have relatively flat bases.

CONVECTION, OROGRAPHIC UPLIFT AND FRONTAL ACTIVITY

Convection occurs when warmed, near-surface air becomes buoyant and moves upward to form "puffy" clouds on the updrafts. Orographic uplift results from wind ascending along a mountain to form clouds on the upwind slopes. Frontal activity occurs when warm air ascends along a frontal boundary over cooler air.

CONVECTION

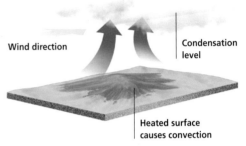

Wind direction

Condensation level

Heated surface causes convection

OROGRAPHIC

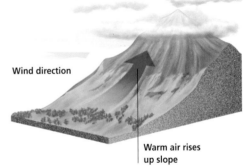

Wind direction

Warm air rises up slope

FRONTAL

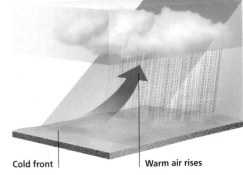

Cold front

Warm air rises

Vortices and waves

Clouds often form over mountainous regions where the prevailing flow of air is forced to rise when it hits the mountain barrier. When air ascends on the upwind side of the mountain, the drop in pressure due to the increasing altitude allows it to cool and expand. As it cools, the rising humid air can become saturated with water vapor, which eventually condenses to form cloud droplets. If the low-level air rising up a mountain slope is sufficiently humid, cloud and fog form along the upwind slopes. The effects of mountains on prevailing winds are often seen as wavelike vertical "motions." As these damped motions increase, they may be detected by a series of wave clouds downwind of the mountains.

Sometimes, clouds formed in a relatively stable atmosphere, such as over the ocean, may produce what appear to be swirling vortices when seen from space. These vortices may originate either when a weak area of low pressure forms or when the overall wind regime encounters a barrier such as a mountainous island that extends through the layer that caps the clouds.

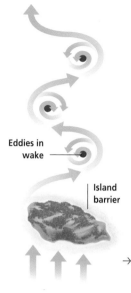

Eddies in wake

Island barrier

Wind direction

HOW ISLANDS BREAK UP CLOUDS
The horizontal flow of air in a stable environment, such as over the ocean, carries clouds around an island barrier. The peaks on the island extend upward through the temperature inversion that caps low-level vertical motion and cloud formation, thereby forcing the flow of air around the island barrier rather than over it. This barrier produces eddies in the wind downwind of the island. When viewed from above, these vortices appear in the cloud pattern in the wake of the barrier. This arrangement, called von Karman vortices, often develops multiple eddies with opposite rotation produced by an oscillating flow downwind of the island barrier.

→ **Born of the mountain** Clouds like this—known as "lenticular" for their oval, lens shape—are frequently seen anchored over a mountain barrier as air flows through the cloud mass.

↓ **Wind over the Canaries** The effect of wind blowing high over the Canary Islands, off the western coast of Africa, is evident in this satellite view. The wind is blowing from left to right, producing turbulent wake patterns downstream of the islands. The existence of these patterns was not fully appreciated until satellite photography revealed them.

→ **Eddies in an island's wake** Each of these swirling clouds is the result of wind-driven clouds encountering a mountain almost a mile (1.6 km) high on Alexander Selkirk Island in the South Pacific, visible on the top left of the photograph. The extreme steepness of the island produces turbulent conditions on the downstream side.

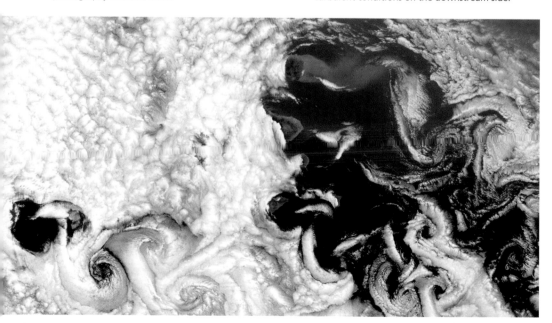

CLOUDS OVER MOUNTAINS

Moist air at mid-levels in the troposphere begins to rise as it moves across the crest of a mountain barrier. During this initial upward motion, a cloud is formed on the upwind side of the mountain as the ascending air expands, cools to saturation with water vapor and condenses to produce cloud droplets. Sometimes "waves" can form in the wind flow.

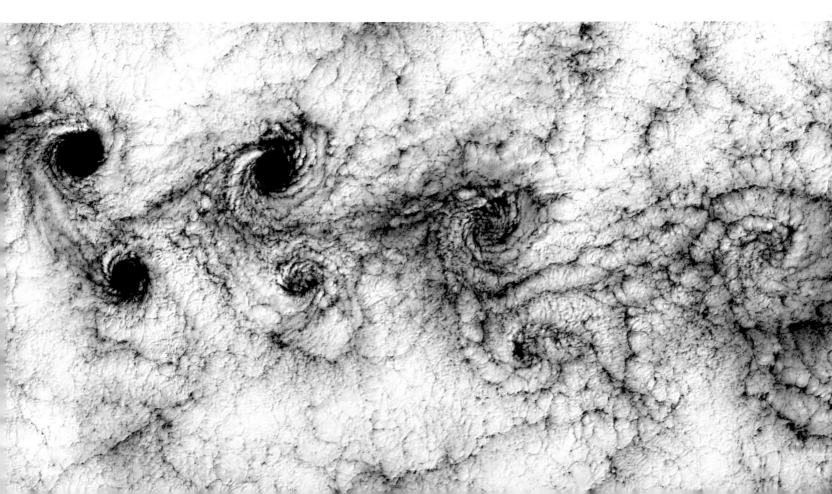

Cloud classification

Since the early nineteenth century, scientists have classified clouds, giving them specific names to aid in their observations. A London pharmacist, Luke Howard (1772–1864), wrote a paper in 1803 entitled "On the Modifications of Clouds," which was the first systematic attempt at classification. Today, the names given to the 10 principal cloud types are essentially a gross description of the main distinguishing features of a cloud in terms of overall appearance and altitude.

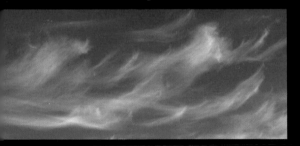

↑ **Cirrus borne on altostratus** The rays of the setting Sun reflect off cirrus and altostratus clouds to dramatic effect. The thin, high-level cirrus contain ice crystals and have a filmy appearance while the thicker, midlevel altostratus clouds contain water droplets and have more definite boundaries.

↑ **As light as a feather** The wispy form of cirrus clouds indicates ice crystals that are being carried along by strong, upper tropospheric winds to form delicate, white shapes that streak across the sky. These clouds typically move at altitudes around 30,000 ft (9000 m).

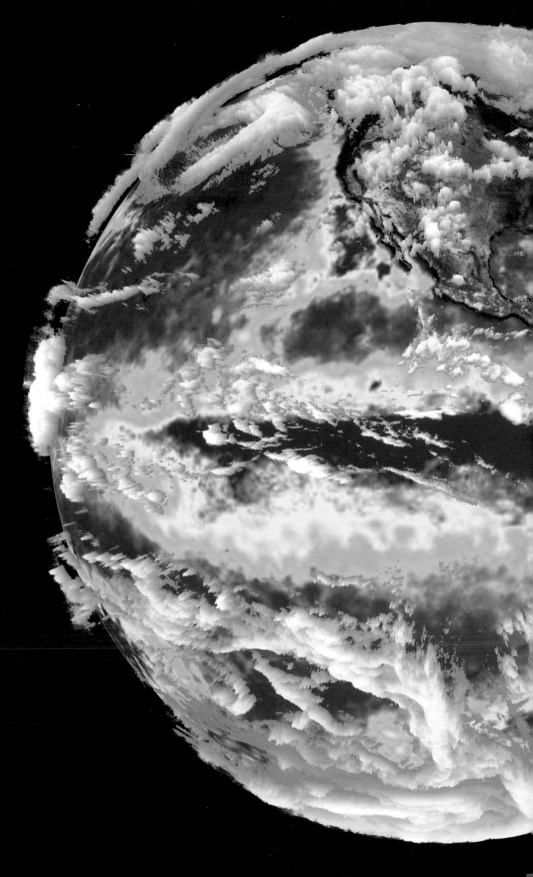

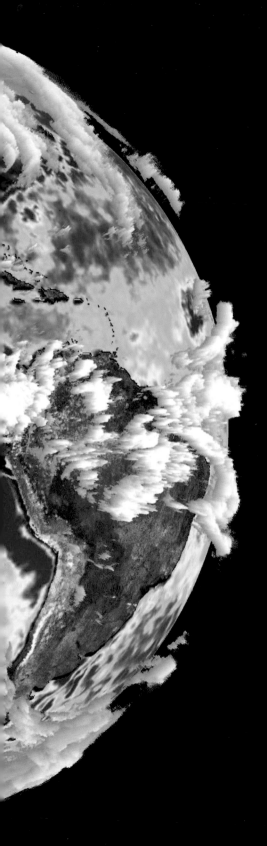

CLOUD CATEGORIES: BASIC TYPES

Alto Derived from the word "high," but in meteorology used to refer to middle-level clouds.

Cirrus Meaning "filament of hair" and used to identify high-level clouds.

Cumulus From a term meaning "pile" or "heap," here refers to a "tall" cloud.

Nimbus Meaning "rain" and refers to rain-bearing clouds.

Stratus Derived from "stratum" or layer, stratus refers to low-level clouds; also used as a suffix to a set of cloud types that have a layered appearance.

CLOUD CATEGORIES: SPECIFIC TYPES

Cirrostratus A combination of cirrus and stratus. Cirrostratus are generally recognizable by a transparent thin white sheet or veil of ice crystals forming high-level clouds.

Cirrocumulus A combination of cirrus and cumulus. High-level ice crystal clouds consisting of a layer of small white puffs or ripples.

Altostratus Stratiform clouds with the "alto" prefix to indicate middle-level altitude. Altrostratus clouds consist primarily of water droplets that appear as a relatively uniform white or gray layered sheet.

Altocumulus A middle-level cloud type that has some vertical development as indicated by the suffix "cumulus." Altocumulus clouds generally have a layered appearance but they also consist of white to gray puffs or waves.

Stratocumulus Low-level layer clouds as suggested by "strato" but having some vertical development as indicated by the suffix "cumulus." Stratocumulus clouds consist of a layer of large rolls or merged puffs.

Cumulonimbus Vertically developed (cumulo) clouds that produce rain, indicated by the suffix "nimbus." These "tall," high clouds usually extend up to the troposphere and have a puffy lower portion and a characteristic smooth or flattened anvil-shaped top. These clouds usually produce heavy rainfall in the form of showers accompanied by thunder.

Nimbostratus Rain-producing (nimbus) layer (stratus) clouds. Nimbostratus are low- to middle-level clouds that have the appearance of a uniform gray layer and which usually have precipitation falling from their base.

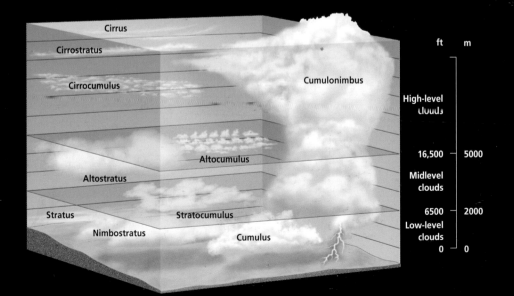

Types of cloud: high clouds

There are three main types of high cloud, all varieties of cirrus—cirrus itself and its two major forms, cirrocumulus and cirrostratus. All three types, usually floating at heights above 20,000 feet (6100 m), are composed of many millions of ice crystals, because temperatures at this altitude are usually well below freezing, and a saturated air mass will produce ice rather than water droplets. Winds are often strong, and help produce the characteristic elongated wispy appearance of cirrus formations. Cirrus clouds sometimes form in isolated patches while at other times they may cover a wide area of the sky. In addition to the three main categories of high cloud, there are also several subdivisions that describe more specialized formations. Among these are the hooklike streamers known as cirrus uncinus and the wavelike formations called cirrus undulatus. When isolated, these clouds may not carry any great significance, but when extensive and occupying a large part of the sky, they can indicate the approach of a frontal system.

STREAKY CIRRUS
Large cirrus clouds may precede an approaching frontal system. An elongated, streaky appearance can indicate strong winds.

Wind blown High-level winds have caused this cirrus cloud to cover a large part of the sky.

ROWS OF CIRRUS
Upper-level "waves" can form when one layer of air slides across another to produce cirrus clouds that form evenly spaced rows.

Familiar pattern These cirrus clouds have lined up in rows, or bands, across the sky.

RAGGED CIRRUS
A large amount of
upper-level moisture
in the atmosphere can
form extensive areas of
spectacular ice-crystal
cirrus formation.

Fragmented cloud
Thick filaments of cloud
have formed this
fragmented cirrus.

SOLID CIRRUS
When cirrus formations
are of a comparatively
solid appearance, it can
mean that upper-level
winds are not as strong
as they usually are.

Moderate winds
Moderate upper-level
winds have resulted in
forked cirrus clouds.

SWIRLING CIRRUS
Cirrus clouds can cover
a large part of the sky in
a haphazard pattern of
swirling skyscapes that
change rapidly if the
wind is strong.

Wavy display
A "disorganized,"
formless cirrus makes
a dramatic display.

ANVIL CIRRUS

The "anvil" can sometimes blow off the top of a decaying thunderstorm to form an extensive area of cirrus downwind.

Post storm This "disorganized" cirrus is probably the remnant of a thunderstorm.

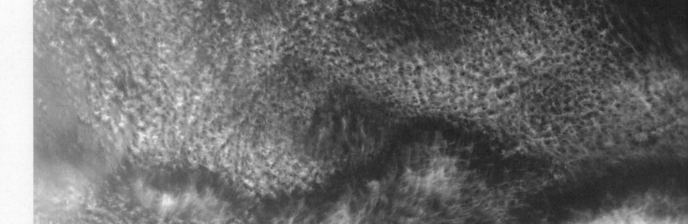

CIRROCUMULUS

Like other members of the cirrus family, cirrocumulus is made of ice crystals; these formations have a cellular appearance.

A million pieces Tiny patches form this attractive cirrocumulus pattern.

SCULPTED BY WIND

Strong winds in the upper troposphere, such as a jet stream, give high cirrus clouds an elongated and streaky appearance.

Painted sky These spectacular cirrus streaks may have been shaped by a jet stream.

CIRRUS SIGN

In midlatitude areas, any marked increase in the amount of cirrus from the west can indicate the approach of a cold frontal system.

Change coming
A cirrus cloud build-up may be a warning of a change in the weather.

UNDULATUS

When one layer of air slides across another at high levels, cirrus cloud can form a wavelike appearance, dubbed undulatus.

Ripples in the sky This cirrocumulus undulatus displays a characteristic rippled appearance.

SUPERB CIRRUS

Large cirrus clouds can form beautiful patterns and are favored by photographers as a backdrop for landscape photography.

Skyscape A sweeping cirrus formation is spectacular on its deep blue background.

CIRRUS UNCINUS
Cirrus formations with a jagged appearance are known as cirrus uncinus—deriving from the Latin word *uncinus*, meaning "hook."

Fiery cirrus A sunset view shows a formation of cirrus uncinus in the western sky.

CIRRUS OVERLAY
These cirrus clouds overlay other clouds at a lower altitude, and indicate significant moisture levels in the atmosphere.

Clouds combined This cirrus formation sits above a lower cloud toward the horizon.

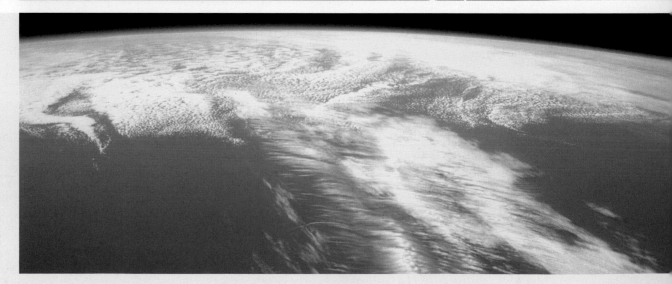

JET-STREAM CIRRUS
Seen from space, jet-stream cirrus formations can be "organized" and extensive. From the ground, however, they may look unremarkable.

From above A satellite view shows jet-stream cirrus over Cape Breton Island, Canada.

CIRRUS TOGETHER

Different cirrus types are frequently seen together, as the conditions for their formation in the upper atmosphere are only slightly different.

Uncinus overlaid
A cirrus uncinus cloud (bottom) is overlaid by a patch of cirrocumulus.

FAINT CIRRUS

When the conditions for formation are only just being met, cirrus—like other cloud types—can appear in small, short-lived groupings.

Passing cloud Small, localized cirrus clouds like this formation may appear for brief periods.

OBEYING THE WIND

Cirrus cloud filaments oriented transversely and forming a band often indicate strong winds blowing in the band's direction.

Transversing These transverse filaments of cirrus are embedded in a band across the sky.

Types of cloud: middle clouds

There are two main types of middle-level cloud: altocumulus and altostratus. Although the prefix "alto" is derived from the Latin word *altus*, meaning high, these clouds are in fact found below cirrus clouds but well above low-level clouds. They float at heights between 6500 feet (2000 m) and 20,000 feet (6100 m), and are normally composed of water droplets, which give them a sharp outline. But they can also be made up of ice crystals, as temperatures at this altitude may fall below the freezing level. Turbulence in the winds in the middle levels of the atmosphere leads to the appearance of wavelike patterns in some forms of these clouds. Several subdivisions of middle-level clouds describe the features of the formations in more detail. Some types, such as altocumulus castellanus and altocumulus floccus, indicate decreasing atmospheric stability combined with increasing moisture, which may be a forerunner to thunderstorm development. Of particular interest is altocumulus lenticularis—a smooth, elongated cloud with a distinctive lens shape.

SIGNS OF RAIN
A thickening layer of altocumulus and altostratus cloud beneath an overcast deck of cirrostratus may mean rain is on the way.

Both types The foreground altocumulus clouds are backed by thin altostratus clouds.

CONDITIONS ALOFT
Round and irregular formations of altocumulus clouds indicate abundant moisture but relatively light winds.

Irregular altos These irregular altocumulus clouds are gradually thickening and may bring rain.

LITTLE MOISTURE
Flat, translucent altocumulus indicate that only shallow layers of moisture are present in a relatively stable atmosphere.

Ripple effect This classical "mackerel" sky altocumulus cloud has irregular ripples.

ALTO SKY
A sky filled with altocumulus with small ripple-like patterns indicates that middle-level winds are of moderate strength.

Alto deck A "deck" of rippled altocumulus displays some slightly thickened elements.

FILLING THE SKY
Large areas of cloud gradually filling the sky mean increasing moisture and perhaps an approaching rain-bearing system.

Wavy alto This dense, rippled altocumulus has formed like waves breaking on a beach.

MOUNTAIN WAVES

Lens-shaped altocumulus lenticularis clouds form over mountains where winds create "waves" and there is a good supply of moisture.

Range clouds These altocumulus lenticularis owe their shape to a neaby mountain range.

CHANGING COLOR

As middle-level winds and moisture increase, rolls of altocumulus clouds thicken, turning gradually from white to gray in color.

Alto rolls Dense rolls of altocumulus indicate increased levels of midlevel moisture.

TURRETS IN THE SKY

Altocumulus castellanus are clouds with "turrets" that warn of developing instability and the possibility of a thunderstorm.

Lit from behind These turreted altocumulus castellanus make for a spectacular sunset.

ALTO LAYERS
Altostratus sheets generally indicate the presence of moisture in the middle levels; if the cloud continues to thicken, rain can follow.

Sunrise This thick deck of altostratus formations is colored by the rising Sun.

RAIN CLOUDS
Continued thickening and lowering of altostratus can lead to steady, uniform rain from an almost totally overcast, dark gray sky.

Sun screen A thick altostratus formation reduces the Sun to a mere glow in the sky.

THREATENING ALTO
When altocumulus forms fluffy heaps—altocumulus floccus—the atmosphere is moist and unstable weather may follow.

Harbinger of rain This large area of altocumulus floccus has filled the dawn sky.

POLAR SKIES

In near-polar regions, thick altostratus layers on the poleward side of intense lows can obscure the sky and bring steady snow.

Sunless sky A dense, overcast layer of altostratus hangs over an icy polar landscape.

CLOUD BREAK-UP

Slight changes in middle-level moisture in a stable weather pattern can lead to clouds dissipating over a relatively short distance.

Contrast A "mackerel" sky of middle–level cloud contrasts vividly with its background.

JET-STREAM CLOUD

Strong jet-stream winds can produce bands of cirrus, cirrocumulus and sometimes altocumulus, even when the lower levels are very dry.

Space view Above the Red Sea, cirrocumulus jet-stream streaks float over some altocumulus.

CLOUDBANKS

Broad-scale atmospheric lift forms altocumulus and altostratus clouds big enough to cover huge areas of the sky at any given time.

Overwhelming These thick, wavy bands of altocumulus dominate the whole sky.

THUNDER CLOUDS

Extensive amounts of altocumulus and altostratus clouds are produced as a byproduct of the formation of thunderstorms.

Chaos above A chaotic sky indicates great instability with possible thunderstorms.

CHANGING CLOUDS

An extensive cover of altocumulus floccus can gradually transform into altocumulus castellanus as the atmosphere becomes more unstable.

Fluffy formation The "turrets" indicate that these altocumulus floccus are changing.

Types of cloud: low clouds

There are five common low-cloud types. Cumulus clouds have a cauliflower-shaped top and a flat base; they generally form when localized pockets of warm air rise. Stratus clouds have a layered appearance, and occur when relatively large areas of moist air rise gently to a level where condensation occurs. A mix of the two types are stratocumulus clouds—layered clouds with convective elements that have very little vertical development. Then there are thunderstorm clouds—the heavy shower-producing cumulonimbus with a fibrous top, often anvil shaped. Finally, there are the heavy rain-producing nimbostratus clouds. These have a base that is generally of a ragged nature. Low clouds typically have bases below 6500 feet (2000 m) and are made up mainly of water droplets, although "tall" clouds with substantial vertical development contain ice and snow and, in cumulonimbus formations, hail. There are minor variations, too: cumulus humilis is broader than it is long, and cumulus mediocris is as tall is it is wide.

HILLS OF CLOUD
Cumulus clouds usually have well-defined but constantly changing cauliflower tops. From an aircraft, they look like undulating hills.

Cauliflower cloud
This sea of cauliflower-topped cumulus extends to the horizon.

SHALLOW CUMULUS
When conditions are stable but there is enough moisture for surface heating to form clouds, shallow, fair-weather cumulus forms.

Drifting by Flat-topped cumulus clouds like these are a common sight on a fine day.

SHROUDING CLOUD
Steep mountain ranges often provide additional lift for the formation of deep cumulus that can seem almost anchored to the mountain.

Cloudy peak This towering cumulus cloud shrouds a snow-clad mountain peak.

STRATIFORM CLOUD
Low-level stratiform clouds form when moisture levels are high, atmospheric conditions are stable and there is little wind.

Stationary cloud This uniform layer of stratus sits suspended above a lake's glassy waters.

STRATOCUMULUS
Vast expanses of ocean are dominated by high-pressure systems that produce a subsidence inversion, forming large areas of stratocumulus.

Cloud slabs From space, this deck of stratocumulus looks like broken ice.

CUMULUS SPREAD

When middle-level temperature inversions are present, cumulus clouds spread out, some of them becoming stratocumulus.

Transformation These dissipating cumulus have spread to form a stratocumulus cover.

CLOUD PATTERNS

With conditions unstable and storms forming, strong downdrafts and updrafts can create tortured but evocative cloud bases.

Sign of change The base of this developing cumulonimbus shows a complex pattern.

BRIEF CUMULUS

Strong winds blowing up the sides of steep mountains often produce ever-changing, billowing cumulus that rapidly dissipate.

Passing formation A formless, ethereal cumulus cloud floats past a rocky crag.

ICY CUMULUS

When the tops of towering cumulus clouds become "fuzzy" it indicates that ice is forming and showers are becomming likely.

Overview A pilot's-eye view of the glaciated tops of a bank of towering cumulus.

CHANGING FORM

Large-scale low-pressure systems can cause clouds of all types to change their form as they spiral toward the center of the system.

Into a spiral This cloud, seen from space, is spiraling into a low-pressure center.

ROTATING CLOUD

Strong horizontal rotation of clouds can occur on the edge of squall lines due to the interaction of updrafts and downdrafts.

Squall ahead A huge squall line shows a stark contrast of white and gray in a single cloud.

COASTAL CLOUD

Subtropical, coastal locations are frequently subjected to extensive stratocumulus cloud sheets that move in from the moist oceans.

Ocean sent This maritime stratocumulus formation is blanketing a coastal city.

MOUNTAIN CLOUD

In mountainous regions with very moist surface conditions, clouds form on the windward side only to dissipate on the lee side.

Windward stratus A windward mountain slope is draped with a thick layer of stratus.

CLOUD INDICATOR

At sunrise, clouds with ragged or rounded tops indicate atmospheric instability; deeper clouds can be expected later in the day.

Reflections These small fractocumulus clouds are mirrored in still, pre-dawn lake waters.

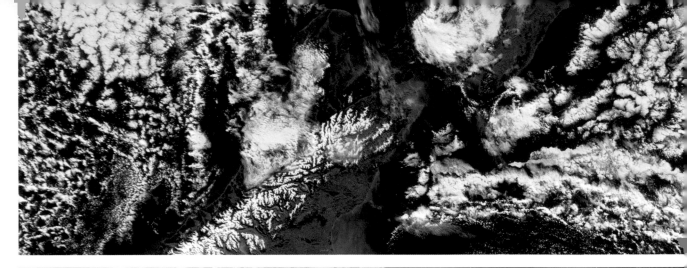

MONITORED CLOUD
Satellite imagery provides meteorologists with a tool with which to continuously monitor the formation and movement of clouds.

Pacific cumulus
A maritime cumulus over the South Pacific near New Zealand.

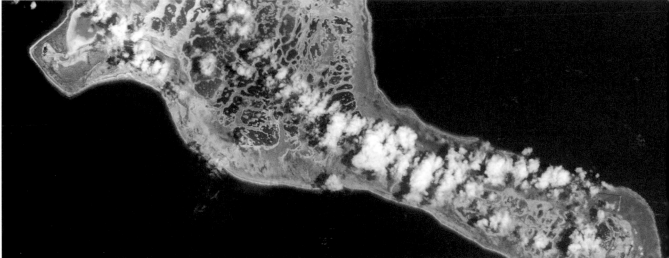

SEA-BREEZE CLOUD
The lift provided by converging sea breezes helps to form towering cumulus slightly inland from the coast of peninsulas and islands.

Island cumulus
A satellite camera captured this cumulus over Christmas Island.

MASSIVE CLOUD
When low-level wind convergence triggers thunderstorms over the ocean, small cumulus can rapidly become huge cumulonimbus.

Skyscraper As dawn breaks, this towering cumulus cloud casts long shadows.

Types of cloud: vertical clouds

Many middle- and high-level cloud formations have great horizontal extent but are not developed vertically—that is, they are not very tall or thick. But there are other types of cloud, usually formed by convection, that attain great vertical development; some of these actually extend from low levels to the top of the troposphere, which is effectively the upper limit for most cloud formation. These giant clouds that literally fill the visible sky are called cumulonimbus or thunderstorm clouds, and can attain heights nearly double that of

Mount Everest, almost 60,000 feet (18,000 m) into the upper atmosphere. Fully formed, they are crowned with a huge, wedgelike anvil-shaped mass of cloud. Considerable vertical development can also occur with smaller clouds such as the so-called towering cumulus that frequently develop into cumulonimbus clouds later in their lifecycle. These vertical clouds are among the most spectacular of all cloud formations; their splendor and beauty make them a popular subject for photographers.

SPREADING CLOUD
Temperature inversion high in the troposphere gives cumulonimbus a horizontal spread, producing their typical anvil shape.

Anvil clouds These cumulonimbus have the anvil shape common to thunderstorms.

THUNDER CLOUDS
As thunderstorm activity develops, some cumulus clouds can form into towering vertical clouds that produce local heavy showers.

Growth potential
Some of these towering cumulus may eventually become cumulonimbus.

STORM ROTATION
Thunderstorms can develop at different rates in the same area, with developing storm cells replacing mature cells as they dissipate.

Storm cell The cumulus clouds around this thunderstorm cell could mature into storms.

LONE CLOUD
Under the right conditions, large thunderstorm cells can develop in isolation, allowing an unimpeded view of a giant cloud.

Single cell Strong winds are tilting the top of this thunderstorm cell toward the right.

RISING COLUMNS
Columns of cumulus cloud can sometimes bubble rapidly upward on their way to building into a cumulonimbus formation.

Bubbly sky This cotton-wool look is the result of strongly growing cumulus "bubbles."

RAIN CLOUDS

When thunderstorms amalgamate, a wide area of land can be overshadowed by cloud; local heavy falls are the usual result.

Stormy weather
A storm cluster "boiling" above Texas, USA, in June 1991.

TURBULENT CLOUD

The turbulence inside a developing cumulus formation can provide a bumpy ride for passengers in aircraft traveling through.

Aerial view This evolving cumulus show a zone of potential storm development.

CLOUD PEAK

The crown of a cumulonimbus cloud is the spectacular, anvil-shaped top, much of which is composed of tiny ice crystals.

High altitude The top of a thunderstorm like this can reach over 30,000 feet (9000 m).

MONSOON CLOUD

In tropical regions, wet-season thunderstorms often occur on a daily basis, usually during the late afternoon or early in the evening.

Storm forming This 1984 space shuttle view shows a storm forming over Brazil.

STORM SEASON

The monsoons of south and southeast Asia produce widespread thunderstorm activity, with frequent heavy rain and flash flooding.

Monsoon The center of this image shows a cluster of monsoonal storms above India.

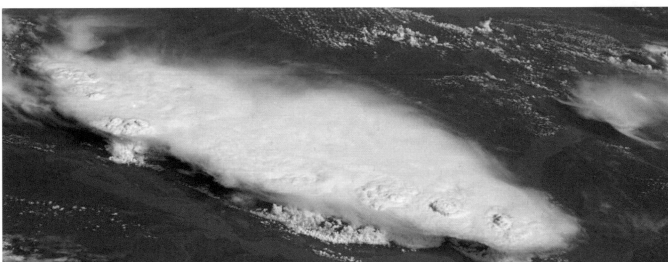

AFTERNOON STORM

Thunderstorms often develop over land in mid-afternoon, which is generally the time of maximum air temperature.

Single cell A single thunderstorm cell develops north of Lake Poopo in Bolivia.

Types of cloud: unusual clouds

There is a kaleidoscopic range of cloud forms, patterns and colors, most of which become familiar to the observer through everyday association. But some specific types are seen more rarely, either because an unusual set of atmospheric conditions is required for their generation, or because they occur in remote areas. These formations are often fascinating to observe. Thunderstorms, too, can produce amazing visual effects, as do some types of high cloud, particularly when the Sun is at a certain angle. Clouds over mountain areas can have a distinctive appearance because of the turbulent effects of winds passing across steep terrain. These clouds often take on regular, rounded shapes and are believed to have been responsible for a number of UFO sightings, particularly at night—as they reflect the light of the moon they have been identified by observers as flying saucers. Views of clouds from space have revealed strange and intricate patterns not obvious to a ground observer. The intricacy and variety of clouds are unsurpassable.

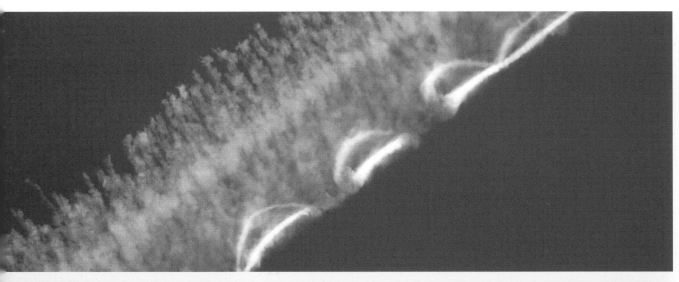

MAN-MADE CLOUD
The exhaust emissions of high-flying aircraft can produce cirrus-type clouds when they interact with below-freezing temperatures.

Jet trail This contrail has become diffuse, spreading out under the influence of the wind.

LINES OF RADIATUS
Strong winds are often the cause of cirrus radiatus forming long lines that seem to radiate from a single part of the horizon.

Natural beauty The setting Sun highlights a spectacular, fan-shaped cirrus radiatus.

UNUSUAL SHAPES
Some cloud formations take on intriguing and complex shapes because of complex, interacting atmospheric processes.

Tornado-like Tubular-shaped clouds like this can develop a tornado-like appearance.

FILTERING CLOUD
When a cloud occurs at the right elevation near sunrise or sunset, so-called crepuscular rays scatter light in the lower atmosphere.

Illusion These dark shadows across the sky are caused by crepuscular rays.

SMOOTH CLOUD
Strong winds that strike a mountain range are pushed upward, resulting in smooth, lens-shaped cloud called altocumulus lenticularis.

Lenticularis A fine example of altocumulus lenticularis has formed over this mountain.

SPIRALING CLOUD
Space photography reveals the contortions of vortex streets or von Karman vortices—eddies of low-level cloud blowing over islands.

Wind impact These von Karman vortices were photographed over the Canary Islands.

ABSTRACT CLOUD
The intricate patterns of some von Karman vortices are shaped by the terrain below as well as the current weather conditions.

Delicate These delicate clouds were shaped by winds over the northern coast of Russia.

DIFFERENT VIEW
Unlike standard camera lenses, which "see" a fairly small sector of the sky, wide-angle lenses provide a dramatically larger field of vision.

Skyscape The wide field of vision of a fish-eye lens produced this spectacular cloud view.

WIND IMPACT
Winds blowing across islands produce von Karman vortices as well as equally dramatic changes in wave formations.

Double impact Winds over the island at right have affected clouds above and waves below.

ISLAND IMPACT
When the appropriate weather conditions prevail, von Karman vortices can occur across just about any island formation.

Vortices These well-developed von Karman vortices are off the eastern coast of Mexico.

CLOUD TRAILS
Complex vortices can trail chains of counter-rotating eddies for many miles downstream of the land that set off their formation.

Patterns These intricate eddies are downstream of the Cape Verde Islands.

Fog and mist

Fog is essentially cloud that forms near the ground and, like cloud, forms as a result of condensation. As it condenses, water vapor in contact with the ground adheres to atmospheric particles such as dust specks. While most fog consists of water droplets, ice fog often forms as a collection of ice crystals in polar regions where temperatures may fall below -22°F (-30°C). Mist, often confused with fog, is actually a suspension of tiny droplets that does not reduce visibility to the same extent. Fog droplets form either by the addition of water vapor or the cooling of air; as a result, there are several types of fog—radiation fog, upslope fog, advection fog and steam fog.

Fog can be eerie and mysterious, or tranquil and calming. As it reduces visibility, it can also be dangerous for motorists, mariners and pilots. Thick fogs can develop in cities, where there are millions of tiny particles on which water vapor can condense. Fog combined with dust or smoke is known as smog.

→ **Golden Gate enshrouded** Moist onshore winds flowing over the cold waters of the Pacific Ocean form advection fog that rolls through the Golden Gate Bridge and over the hills into San Francisco Bay, USA, during the summer months.

↘ **Down in the valley** Fog is often associated with valleys—it is usually the result of overnight radiational cooling on already cool air that has drained to the lowest place in the landscape, deep in a valley.

↓ **Floating fog** Radiation fog forms in valleys with clear skies above and a fresh snow cover below. Essentially, radiation fog is cloud on the ground. It forms when heat from Earth's surface is radiated out to space and the ground cools. This cools a layer of air near the ground, where moisture condenses into millions of droplets.

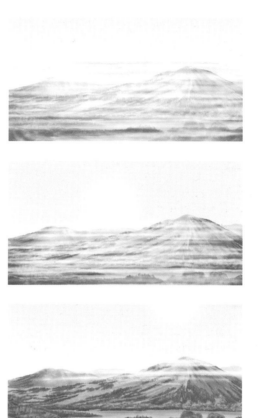

LIFTING OF THE FOG

The Sun's warming effect erodes a fog bank from the edges, causing it to lift gradually. Night fog begins to erode as the morning Sun rises above the horizon, heating the ground and air. Heating the air reduces the relative humidity and also helps "stir" the air, bringing down drier air from above, and causing the fog base to rise. In most cases, these processes mean that by midday the fog has disappeared.

If fog is particularly deep and thick, the Sun's rays may not penetrate it to warm the surface, and the fog will not lift. Cold days result when a foggy morning is followed by the arrival of a cloudbank that blocks the Sun.

Drifting over the channel A computer-enhanced satellite image provides a clear image of an extensive bank of sea fog—common in spring and summer—to the north and northeast of the British Isles.

Kinds of precipitation

Precipitation is what keeps water moving constantly from its various
"reservoirs"—from the atmosphere to oceans, rivers, glaciers and ice
caps. By definition, precipitation is any liquid water or ice that falls
freely from the atmosphere to Earth's surface under the influence
of gravity. Three types of precipitation are recognized: liquid, solid
and freezing.While many clouds may form, relatively few produce
significant precipitation. Generally speaking, clouds thicker than
4000 feet (1200 m) are able to produce precipitation, with only
nimbostratus and cumulonimbus clouds being responsible for most
heavy precipitation. In other kinds of cloud, the cloud droplet or ice
crystal is often too small to fall through the updraft in the cloud and
reach Earth's surface as precipitation. Even if the raindrop were
sufficiently large, it may evaporate below the cloud base.

↗ **Held in an icy grip** Rain falling onto a cold tree branch will freeze on contact if
a thin layer of air surrounding the branch is below freezing point. In an ice storm,
ice accumulation can become so heavy that it breaks large branches.

→ **Wet precipitation in the tropics** Precipitation as rain and drizzle is common in
the Indian city of Darjeeling during the monsoon season. Rain forms in a cloud
as either water droplets or ice crystals. The type of precipitation that reaches the
ground depends upon the temperature structure of the lower atmosphere
between the cloud and the ground.

→ **In the path of the drifting snow** Wet snow with high water content and
falling at temperatures near freezing point often sticks to exposed objects such
as tree trunks and branches. Some trees, such as conifers, have evolved a sloping
shape that reduces snow accumulation to the minimum.

RAIN AND SNOW

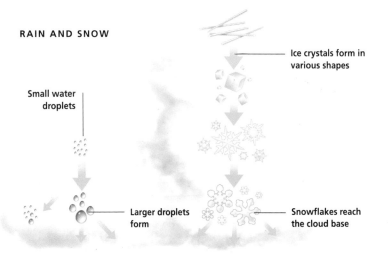

Ice crystals form in
various shapes

Small water
droplets

Larger droplets
form

Snowflakes reach
the cloud base

KINDS OF PRECIPITATION

Meteorologists group precipitation into liquid (rain and drizzle), solid (snow, ice pellets and hail) and freezing (freezing rain and freezing drizzle). Rain is liquid precipitation that falls primarily from nimbostratus or cumulonimbus clouds with drops at least 0.02 inches (0.5 mm) in diameter. Drizzle is numerous small liquid drops with diameters between 0.01 and 0.02 inches (0.25 and 0.5 mm) that come from stratus clouds. Snow is frozen precipitation that consists of white ice crystals arranged in a variety of branched and hexagonal forms, often forming snowflakes. Precipitation can also be classified as either steady or intermittent. Steady rain or snow usually results from frontal activity, while intermittent precipitation is a combination of convection and atmospheric instability.

← **Worldwide phenomenon** Japanese artist Ando Hiroshige (1797–1858) painted *Sudden Shower at Ohashi Bridge at Ataka*. Driving rain like this can fall from stratiform clouds. These clouds normally cover a wide area, so rainfall tends to be widespread and long lasting. Showers are falls from cumuliform clouds; they are more localized and shorter-lived. Rain occurs worldwide except in the polar regions, where precipitation falls as snow.

↑ **Urban whiteout** New York traffic is reduced to a crawl when precipitation falls as heavy snow—the result of temperatures in the lower atmosphere being at or near freezing point. At such temperatures, snow becomes likely when there are high levels of moisture in the atmosphere.

Rain and drizzle

Rain is liquid precipitation that falls to the ground from clouds. Rain-producing clouds are sufficiently deep to develop large raindrops. Much smaller drops may fall as drizzle from some stratus or cumulus clouds that are not deep enough to produce heavy raindrops. Drizzle usually appears as a fine mist that falls or drifts so slowly through the air that it can appear almost suspended in the atmosphere. Even if drizzle persists for an extended period, its impact is so slight that only small rainfall totals are recorded on the ground. Rain, on the other hand, can range from very light to extremely heavy, producing widespread floods and disruption to everyday life. Despite this, rain is crucial to the water cycle: it is one of the main ways by which water in the atmosphere returns to the oceans and is vital in replenishing Earth's reserves of freshwater.

→ **Nature's propagating tool** Apart from being a basic requirement for plant life, rain also helps in the propagation of some species. In this close-up photograph, the impact of the raindrops on the cup-shaped fungus causes its tiny spores to "splash," like water droplets, into the air.

↘ **Urban motorist's nightmare** Rain may be a blessing for farmers but for the city dweller it can be a major inconvenience, slowing traffic on highways and at airports. The run-off from paved roads and sidewalks after a heavy downpour often results in localized flooding.

↓ **Inventive solution** These Indonesian schoolgirls have stowed their shoes in an upturned umbrella as they cross a flooded street in Jakarta during a flash flood in January 2002. Flash floods such as this are common throughout the tropics during the monsoon season.

"SUN" SHOWER

At times, rain can fall when there are no clouds directly overhead. The wind can carry the falling rain out from under the parent cloud, giving ground-level observers an unexpected shower. This phenomenon is thought by some to be the origin of the popular saying about things happening "out of the blue," meaning an unexpected event or occurrence. By contrast, freezing rain is common in regions that experience frequent winter snows. When temperatures at cloud level are below zero, water droplets that fall from clouds will be supercooled and are likely to freeze when they encounter a colder layer of air or a surface below freezing point. Precipitation that freezes in either of these ways is known as freezing rain. The rain may turn into tiny pellets in mid-air and fall as sleet (in the United States) or ice pellets (in the United Kingdom and Australia).

← **Not what it seems** Sometimes rain fails to appear even when clouds are directly overhead. A layer of dry air beneath a cloud can cause the raindrops to evaporate as they pass through it. This phenomenon, often confusing to a person on the ground, is called virga. Because virga does not reach the ground, it cannot technically be classified as precipitation. However, the evaporation that produces virga increases the water vapor content in the layer of dry air and thus makes it more likely that subsequent falls will reach the ground.

↑ **Snap frozen** A shallow layer of subfreezing air close to ground level can cause rain to freeze on contact with objects such as twigs and flowers, temporarily encasing them in a layer of ice.

DROPS, BIG AND SMALL

A comparative illustration shows that raindrops (*below left*) are much larger than drizzle drops (*below, second from left*). A typical raindrop has a diameter of about $\frac{1}{50}$ inch (0.5 mm), whereas an average drizzle drop is usually no more than $\frac{1}{125}$ inch (0.2 mm) across. Because of its larger size, when a raindrop is released from a cloud it falls faster than a drizzle drop; air resistance slows the tiny drizzle drop to a very slow speed. This slow descent explains why drizzle may seems to drift rather than fall directly to the ground.

↓ **Drops, spherical and spheroid** Small raindrops adopt an almost spherical shape (*below, second from right*) and large raindrops attain an oblate spheroid shape with a flattened bottom due to air resistance as they fall (*below right*).

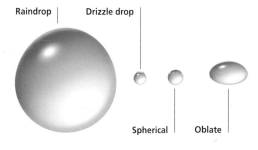

Raindrop | Drizzle drop

Spherical | Oblate

Snow, ice and hail

Snow, hail and ice pellets fall when a sufficiently thick layer of air near Earth's surface is at or below the freezing point of water. During winter, snow in the form of snowflakes can fall from stratus- or cumulus-type clouds. Hail is a product of thunderstorms, where ice particles form when supercooled water droplets pass repeatedly through cloud layers of different temperatures. Ice pellets are produced when rain falling through a warm layer of air encounters a deep layer of subfreezing air that turns them solid. Occasionally, ice pellets may accumulate on the ground.

In many regions, a buildup of snow on the ground is important for agriculture and water storage, since a snow cover helps to slow percolation of meltwater into the soil and reduce water loss. In mountain areas of the western United States, for example, the snow pack and accompanying snowmelt provide a major portion of the water needed for irrigation, energy generation and industrial uses.

→ **Snow . . . but only on the lake** During winter, cold air moving across the relatively warm waters of Lake Superior, the northernmost of the Great Lakes, on the Canada–US border, contributes to lake-effect snow. The air becomes saturated and convective clouds form, accompanied by squalls and snow grains.

↘ **Instant ice** Rain landing on a surface with a temperature below 32°F (0°C) freezes on contact, forming an icy glaze. Freezing rain showers can become serious ice storms in which accumulations of this glaze damage overhead utility lines.

↓ **Proceed with care** Snow falling in the built-up environments of cities and towns ranges from being a minor inconvenience that causes traffic snarls to a serious hazard associated with vehicle accidents and infrastructure failure.

Sun-tinted skyscape The highest part of this cloud receives the most light and so still appears white; the lower parts receive less and take on a coppery glow.

The unmatchable drama of a sunset A cloudy horizon and dust in the atmosphere that scatters light can produce spectacular colors at sunset. Rays of sunlight passing around well-defined clouds are called sunrays or crepuscular rays.

WHY THE SKY IS BLUE

Sunlight travels through the atmosphere in straight, invisible waves. This so-called white light is a mixture of all the colors of the spectrum: red, orange, yellow, green, blue, indigo and violet. Each color of the visible spectrum travels at a different wavelength, with red and orange having the longest wavelength, and indigo and violet the shortest. The color of rays of sunlight changes between midday and just before sunset. The diagram at right depicts a high Sun angle, such as near local noon, when polychromatic sunlight passes through the atmosphere along a relatively short path length. The gaseous molecules in the atmosphere scatter the sunlight, starting with the violet end of the spectrum. When the Sun is high in the sky, only the violet, indigo, blue and a little green are scattered, producing a blue sky.

The diagram at bottom right depicts sunlight passing along a long path through the atmosphere near local sunset. At this time the path length may be as much as 30 times longer than at noon. With significantly more molecules in the path followed at this time of day, more blue, green and even yellow light is scattering into the sky, effectively leaving only the red and orange light to reach the observer. A similar pattern occurs at sunrise. The red and orange colors of sunset can be intensified by air pollution, ash and smoke from fires, and even volcanic eruptions many thousands of miles away.

The way in which water droplets in clouds scatter light creates the white color of the clouds.

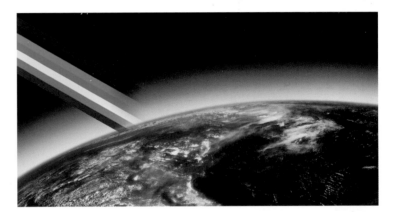

Rainbows and coronas

Light from the Sun passing through airborne cloud droplets or falling raindrops can produce interesting and often beautiful optical effects. If the observer faces away from the Sun and sees a rainbow in falling raindrops, the Sun's rays have traveled along a complicated path. As the light passes through the drops, it is refracted or bent and dispersed into its component colors. Some of the light that has entered the drop is reflected, permitting the observer to see the different colors as they form a rainbow. But if the observer looks toward the Sun through a thin layer of clouds containing relatively uniformly sized water droplets, a corona and an iridescent color may be detected. As the light rays pass around the tiny spherical droplets, interference or diffraction patterns with alternating colored bands develop. A corona is best seen when the Moon, rather than the Sun, is the light source; not only is it inadvisable to look directly into the Sun but its brightness tends to swamp the coronal effect. Iridescence appears as irregular patches of color in midlevel clouds adjacent to the Sun or Moon. It is, in fact, an imperfect corona as it is formed by the same process of light diffraction around water droplets. It lacks the symmetry of a corona, appearing as diffuse patches or bands of color.

EXPLAINING THE PHENOMENON
Rainbows have fired imaginations throughout human history, often assuming great religious significance as they periodically returned to the sky. It was not until the late seventeenth century that a scientific explanation for this phenomenon was provided. Isaac Newton showed that when a beam of light passed through a glass prism it was refracted, breaking down into a spectrum of colors. From this, he correctly inferred that white light was a combination of all the colors in the visible spectrum.

↑ **Dazzling but ephemeral** A combination of Sun and rain are necessary for the creation of a rainbow. The higher the Sun, the flatter the arc of the rainbow.

→ **Shimmering spectrum on the sea** Iridescence produced by the Sun glints off the surface of the ocean. The rainbow-like bands of colors that appear on the sea are caused by the interference or diffraction patterns of the Sun's rays reflected off the water's surface. Most of the colors in the spectrum are visible. Iridescence is sometimes associated with a developing frontal system.

↓ **The unique beauty of a corona** The lightly colored bands that surround the Sun or Moon when viewed through thin altocumulus clouds are called a corona.

↙ **Double rainbow** A complete primary rainbow is shown with portions of a fainter secondary rainbow. The inner band on the main rainbow is blue, while the color sequence is reversed on the second rainbow.

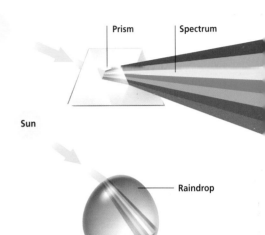

RECIPE FOR A RAINBOW

White sunlight, a mixture of multiple colored light, passes through a prism—such as some form of water—and is bent, dispersing into its component colors (*top*). Sunlight passing into and out of a spherical raindrop is bent and dispersed twice into its component colors (*center*). If the Sun is within approximately 42 degrees of the horizon and there is nearby falling rain, an observer facing away from the Sun and toward the rain will see a rainbow with a series of colored bands ranging from blue on the inside to red on the outside (*bottom*).

Haloes and sundogs

Airborne ice crystals that act like tiny prisms are responsible for the intriguing optical effects known as haloes and sundogs. When sunlight or moonlight passes through the crystals, they cause the light to be refracted or bent slightly. Because many of the ice crystals may be in random orientation, circular haloes result. At certain points on the halo, particularly bright patches form, known as sundogs. Since the bending of light is also dependent upon color, haloes and sundogs exhibit considerable variation in coloring. Ice crystals are not the only agents that refract sunlight, as sunlight is bent as it passes through varying elements in the atmosphere, producing mirages and on occasion, the dust-related phenomenon known as the green flash that occurs just as the Sun disappears below the horizon.

HALOES, A HARBINGER OF RAIN—SOMETIMES

For centuries haloes have been a traditional indicator of the coming of rain. Rain does sometimes follow the sighting of a halo—the cirrus cloud that provides the ice crystals for the refraction that forms a halo sometimes precedes a frontal system that brings rain. However, there are many exceptions to this combination of phenomena, and a halo cannot be regarded as a reliable forerunner of rain. Haloes occur worldwide, but are more common in high latitudes. They vary considerably in size and color, and are sometimes incomplete, forming an arc rather than a full ring.

→ **Polar phenomenon** The Sun, with a portion of a halo along with a pair of brightly flashing sundogs, dominates the Antarctic sky. The phenomenon occurs frequently in polar regions but can also be seen in lower latitudes. Sundogs, also known as parhelia or mock suns, create an eerie illusion of three Suns in the sky.

↓ **Nothing but thin air** The buildings below are a trick of the light—a mirage. When Earth's surface is hot, the increased density of the overlying air can cause light to be refracted back upward, resulting in an image on the horizon.

Shimmering pillar of the Sun This dramatic optical effect, called a sun pillar, is caused by the reflection of the Sun's rays from ice crystals that are drifting Earthward. The crystals originate in cirrus cloud or ice fog floating near the ground. Sun pillars are most likely to be seen in high latitudes, where ice fog is more common.

← **When the Sun turns green** A green flash represents a sudden change in the color of a tiny crescent of the Sun at sunrise or sunset. This small burst of light, changing from red to green, lasts for just a few seconds.

→ **The Sun, framed by light** The Sun's halo is caused when its rays are refracted, or bent, by tiny ice crystals.

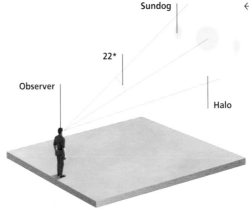

Sundog

22*

Observer

Halo

← **Fleeting effects that delight the eye** Light rays passing through a veil of ice crystals produce halo and sundog phenomena. The most common angle of refraction through ice crystals is 22 degrees, and most haloes are produced in this way. This circular luminous ring results from the relatively random orientation of the tiny crystals.

Auroras

The Sun's surface is volatile. Disturbances sometimes create eruptions of ionized particles, some of which are ejected into space at massive speeds, taking about 30 hours to travel from the Sun to Earth. Earth's magnetic field deflects these charged particles toward the poles, where they create massive magnetic storms. When the particles hit molecules and atoms in the ionosphere, they vibrate; when they return to their original state, light is emitted. As this can happen over a large region, the polar skies become illuminated with displays of colored or white sheets and beams of light—the auroras. They occur between 50 and 600 miles (80–1000 km) above Earth: aurora borealis in the northern hemisphere, and aurora australis in the south.

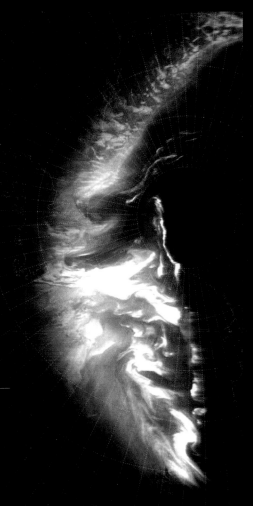

← **An extensive aurora** australis surrounds the South Pole, located at the center of the innermost circle. This photograph was taken 550 miles (887 km) above Antarctica; the latitude and longitûde lines, and the Antarctic coastline, have been added by computer. Auroras appear most frequently around the equinoxes, and at times of maximum sunspot activity. Aurora australis has been seen as far north as Brisbane, Australia, and aurora borealis as far south as Athens, Greece.

→ **A time exposure** from space shuttle Endeavour shows the aurora borealis, or northern lights, as a green glow across the sky. The lights of an unidentified city are in the foreground. Green–yellow auroras occur when electrons collide with oxygen molecules in low-pressure atmospheric regions.

The 1991 space shuttle mission captured these spectacular views of the aurora australis. They show the high-level red aurora and the lower green aurora. Auroral displays take on many shapes: bands, rays, arcs or luminous curtains of multicolored light.

Previous page The funnel of a tornado is graphically
outlined as it sweeps through the American Midwest.

Extreme weather

Weather is perhaps the last wild thing on Earth. The extreme winds of hurricanes, tornadoes, blizzards and ice storms; the destructive power of avalanches, floods, wildfires and droughts—all are reminders of the elemental force of weather.

Thunderstorms

Thunderstorms, one of nature's most awesome phenomena, occur in about 40,000 locations across the world every day. The majority of these storms happen in spring and summer in tropical and subtropical areas; they are absent only in Antarctica. Thunderstorms are formed when cumulus clouds continue to grow until they extend throughout the troposphere, forming mountains of moisture that can reach up to 50,000 feet (15 km) in height. The conditions required to produce this phenomenal cloud growth can be provided by a cold front, when the atmosphere is lifted by a wedge of cold air driving in under the existing airmass. So-called unstable conditions, in which the temperature of the atmosphere decreases rapidly with height, can also result in the development of thunderstorms. A typical thunderstorm will last from one to two hours, before it is slowed by downdrafts that are assisted by the accompanying rain. Occasionally more intense, severe storms last much longer than two hours. They may produce intense bursts of lightning and thunder, heavy rain, hail and strong winds; their impact can be devastating, particularly in urban environments.

↓ **Eye in the sky** In April 1984, cameras on the space shuttle provided this bird's-eye-view of a series of thunderstorms over Florida, USA. Despite the great height from which the photograph was taken, the enormous depth of the thunderclouds is evident.

→ **Hovering tempest** The brooding power always suggested in the buildup of a thunderstorm is obvious from this spectacular cloudscape. Heavy rain is falling across the horizon to the right. After the storm, only a few wisps of cloud will remain.

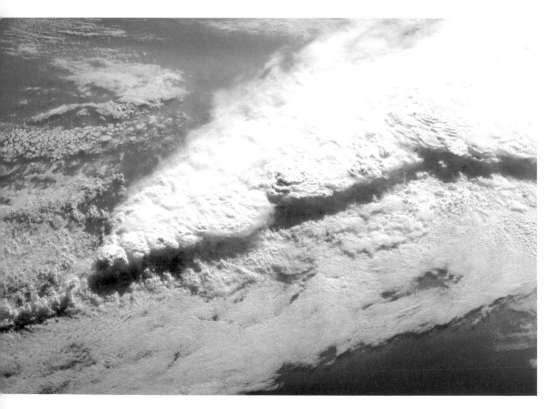

Wild weather The drama of a violent thunderstorm is captured in this photograph of a huge cumulonimbus cloud illuminated by lightning. Most thunderstorms have a three-phase lifecycle. The cumulus phase occurs as cumulus clouds grow, sometimes to massive sizes as shown here. Strong updrafts prevent rain from falling, and there is no lightning. The second phase is the mature stage, when ice particles grow in the upper cloud and become sufficiently large to create precipitation. Downdrafts form; the air becomes colder, more turbulent and electrically charged. Lightning occurs and rain, or perhaps hail, falls. In the third phase, the storm dissipates as the precipitation creates weak downdrafts that deprive the cloud of its energy supply. The cloud evaporates and the storm subsides. This final stage may last for up to an hour. Thunderstorms are sometimes arranged along a line of low pressure, known as a squall line because the downdrafts cause gusty winds at the surface. A fully formed squall line is constantly regenerated by the cooler downdrafts that lift warmer, moister air in its path.

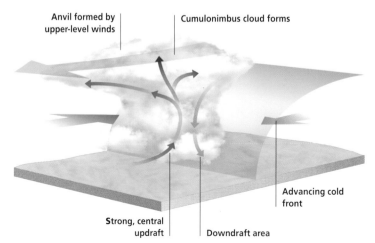

Anvil formed by upper-level winds

Cumulonimbus cloud forms

Strong, central updraft

Downdraft area

Advancing cold front

BIRTH OF A STORM

This diagram shows how a cold front contributes to a thunderstorm. The wedge of cold air associated with the advancing cold front drives under the existing air mass, producing an upward motion in the air. As the thunderstorm develops, updrafts and downdrafts form, and the cloud flattens as it reaches the top of the troposphere. The higher levels of the cloud are then sculpted into an anvil shape.

Lightning

The most obvious features of thunderstorms are, of course, lightning and thunder. Intriguingly, we still do not know exactly what causes lightning, but several facts point to likely reasons for this spectacular phenomenon. It seems that areas of opposite electrical charge build up within cumulonimbus clouds, with a positive charge tending to gather along cloud tops and a negative charge nearer the base of the cloud. Because air is a poor conductor of electricity, these charges continue to accumulate until enormous electrical differences are generated. The imbalance in the electrical charges is corrected abruptly by a gigantic discharge—lightning. Lightning heats the air to over 54,000°F (30,000°C), producing an explosive expansion of air—thunder. The sight and sound of thunder and lightning can be truly dramatic.

CALCULATING DISTANCE

We hear thunder as a loud crack if it is close or as a low, rumbling sound if it is more distant. Because light travels at 186,282 miles per second (299,792 km/s), we see a lightning flash almost as soon as it occurs. But because thunder takes five seconds to travel one mile (3 seconds/km), it is often many seconds before we hear its sound. To estimate how far away lightning is, count the number of seconds between seeing the lightning and hearing the thunder. Then divide the number of seconds by five to calculate the distance in miles (or three for kilometers).

WHEN GOLF BECOMES A HIGH-RISK SPORT
One of the best-known incidents of lightning striking a person involved the professional golfer Lee Trevino. On June 27, 1975, playing at Butler National Golf Club in Chicago, USA, Trevino was struck by lightning—a freak accident that permanently damaged the flexibility and sensitivity of his lower back. He later remarked famously: "If you are caught on a golf course during a storm and are afraid of lightning, hold up a 1-iron. Not even God can hit a 1-iron." Nowadays, organizers of professional golf tournaments often contact their local weather service to request a weather watch for the period of the tournament. If thunderstorms appear on the radar, meteorologists inform the organizers, who may postpone play until the storms have passed.

↑ **The elements unleashed** Lightning strikes in urban areas can produce sudden power surges that disable electrical supplies. This problem has increased in recent times, with electronic equipment such as computer systems particularly vulnerable.

← **Lightning over the ocean** Inky skies are rent by bolts of electricity with towering cumulonimbus clouds in the troposphere. A lightning bolt typically discharges about 100 million volts of electricity and creates a massive increase in temperature.

↑ **Lighting up the sky** This sunset view of a thunderstorm shows a spectacular series of lightning strikes, some reaching the ground, while others dissipate in the surrounding atmosphere. The luminous area toward the middle of the cloud mass indicates that the main lightning flash may have traveled several miles through the atmosphere in a generally horizontal direction before discharging into the ground. The flattish top and anvil shape that usually distinguish cumulonimbus clouds can be discerned readily. The color of the lightning flash indicates the nature of the surrounding air: the flash is red if there is rain in the cloud; blue if there is hail; yellow if the air is dusty; and white if humidity is low. This latter is therefore most likely to generate fires when it hits the ground.

→ **A buildup of sound and fury** As a cumulonimbus cloud develops its characteristic anvil shape, an electrical charge begins to accumulate. As the diagram at right shows, positive charge tends to gather in the upper levels of the cloud and negative charge in the lower. The buildup of these charges and their eventual dramatic imbalance generate lightning and thunder. Lightning tends to strike where the positive charge is greatest on the ground below the cloud. This may be a tall object such as a tree or high building, or, more favorably, a lightning conductor such as a metal rod. The popular belief that lightning never strikes twice is untrue. Skyscrapers can be struck several times a year, and the Empire State Building in New York was once famously hit 15 times in 15 minutes.

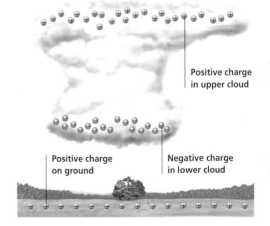

Positive charge in upper cloud

Positive charge on ground

Negative charge in lower cloud

Types of lightning

When positive and negative electrical charges within a cumulonimbus cloud develop to a sufficiently high level, a powerful discharge—lightning—occurs between them. These charges allow lightning to "jump" from a cloud to the adjacent atmosphere; from one cloud to another; or—the phenomenon we are most familiar with—from the cloud to the ground. Lightning that strikes Earth is, of course, the most dangerous to humans; although worldwide, it is most common in the tropics and midlatitudes.

CLOUD-TO-AIR

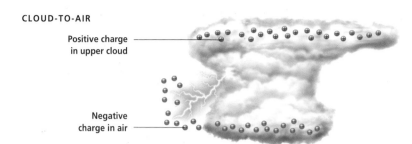

Positive charge in upper cloud

Negative charge in air

↗ **Cloud-to-air lightning** Positive charge accumulating along the higher layers of this cloud creates a difference in electrical "potential" with a negatively charged area in the adjacent atmosphere. Lightning eventually jumps between them.

→ **Cloud-to-cloud lightning** Areas of large difference in electric charge can develop when positive charge along cloud tops is generated close to negative charge near the base of adjacent clouds. Lightning will sometimes discharge across these areas.

↓ **Cloud-to-ground lightning** Strong negative charge along the base of cumulonimbus clouds can induce equally strong positive charge on the ground below the cloud. A discharge eventually occurs between them in the form of a lightning strike.

CLOUD-TO-CLOUD

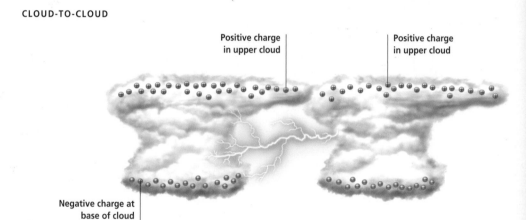

Positive charge in upper cloud

Positive charge in upper cloud

Negative charge at base of cloud

CLOUD-TO-GROUND

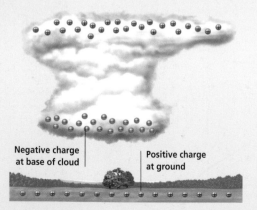

Negative charge at base of cloud

Positive charge at ground

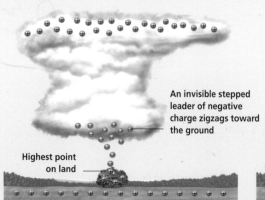

Highest point on land

An invisible stepped leader of negative charge zigzags toward the ground

The circuit is completed in a lightning flash

INTO THE AIR

Lightning may discharge and disperse directly into the surrounding atmosphere. On most occasions, the lightning extends downward from the upper levels of the cloud where the areas of positive charge are located, into a negatively charged area in the atmosphere below the cloud.

Lighting up the sky Lightning from a cumulonimbus cloud—its top obscured by another cloud—is here striking into the adjacent atmosphere.

INTO ANOTHER CLOUD

Cloud-to-cloud lightning can occur between positive and negative charges contained within a single cloud or between several different thunderstorm clouds. The phenomenon of the lightning flash occurring within a single cloud can create a spectacular lighting display, particularly at night.

Inner glow Positive and negative charges within this cumulonimbus cloud have produced a prolonged lightning discharge which has illuminated it from within, resulting in a luminous effect.

INTO THE GROUND

The dramatic flashes usually referred to as lightning bolts are more complicated than they look. The first part of such discharges begins as a generally invisible stepped leader from the cloud to the ground. A return stroke then flashes upward from the ground along the same path as the leader, and several of these up-and-down strokes may occur in one bolt of lightning. Usually this happens so rapidly that the human eye can detect only a single brilliant flash. Each lightning strike lasts only a fraction of a second.

An awesome display A number of spectacular lightning strikes extend from a bank of clouds all the way to the ground. As lightning is potentially dangerous, such displays are best observed indoors.

Hailstorms and microbursts

Severe thunderstorms generate hailstones and wild, gusty winds. Hailstorms are most common in the midlatitudes, particularly in spring and summer. While most hailstones are the size of a pea, some grow as large as golf balls or even oranges. In 1888, baseball-sized hail in northern India reportedly killed 250 people.

Microbursts are thunderstorm-created winds: blasts of air that reach the ground from thunderstorm downdrafts, gusting in excess of 100 miles an hour (160 km/h). Wet microbursts are accompanied by rain and may be almost invisible. Dry microbursts occur in clear air beneath the cloud base and can also be hard to detect. Areas of dust spreading out from beneath the cloud may be the only sign. Microbursts are a major hazard to airplanes landing or taking off.

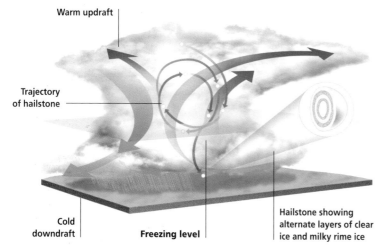

Warm updraft

Trajectory of hailstone

Cold downdraft

Freezing level

Hailstone showing alternate layers of clear ice and milky rime ice

← **Hailstorm waiting to happen** Despite its pale and placid appearance, this is a well-developed thunderstorm cloud with a prominent anvil top composed of ice crystals.

→ **An icy bombardment** This golf-ball-sized hail is the product of a severe thunderstorm. Such large hailstones are capable of causing serious injury to people and animals alike.

HOW ICE FORMS IN THE SKY

Large hailstones are probably the most destructive product of a severe thunderstorm—they can be large enough to damage automobiles, smash through roofs and, on some occasions, even injure livestock and people. Hail begins high in a thunderstorm as ice crystals, which grow as more ice forms about each "stone." After falling within the cloud, the ice particles are swept aloft again in the updraft, and gather more ice. After several such cycles, large hailstones composed of different layers of ice eventually fall to the ground. The size of a hailstone depends upon the number of layers of ice it contains, which itself is based on how long the hailstone remains within the thunderstorm— hailstones consisting of 25 ice layers have been recorded. Hailstorms are usually short lived.

A sudden downpour A sign at the Wimbledon tennis complex in London, England, warns of the dangers that may result from a sudden downpour. Hail-producing storms are most common in the midlatitudes. They damage property, particularly cars, devastate crops and are an aviation hazard.

↑ **Cumulonimbus shower** A shaft of rain falls from a cumulonimbus cloud. These clouds contain a mixture of liquid water droplets that may be supercooled and ice crystals at the top. The crystals melt to become hailstones or, ultimately, rain.

→ **Formation of a dry microburst** A microburst typically begins 3 miles (5 km) above the ground. Precipitation falls inside a towering cloud, dragging nearby air with it. As the air hits the ground, it spreads rapidly from its touchdown point, bringing a burst of very strong winds. Rain evaporating in mid-air forms a dark fringe known as virga. Wet microbursts are formed in the same way as dry, but in this case, precipitation reaches the ground.

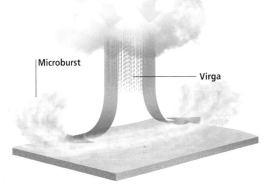

Microburst

Virga

Blizzards and ice storms

During the winter months, blizzards and ice storms occur over large areas of Europe and North America. They can seriously disrupt everyday life by paralyzing transport and creating a dramatic increase in the demand for power, as well as exposing those who have to venture outside to dangerous conditions.

A blizzard is a storm with strong winds—35 miles per hour (56 km/h) or more—and heavy falls of snow, along with very cold temperatures. The combination of these elements creates blowing snow, with near zero visibility, deep snow drifts, and potentially lethal wind-chill. An ice storm is generated when rain falls from a layer of air warmer than 32°F (0°C)—freezing point—into a layer of air close to the ground with a temperature below 32°F (0°C). This can cause pellets of ice—called sleet—to form, and sometimes results in all outdoor surfaces being coated with a layer of ice. Ice storms make any outdoor activity hazardous: sidewalks become slippery, and treacherously icy roads usually result in a spate of highway accidents.

WHEN THE RAIN FREEZES

Ice storms may produce spectacular, picture-postcard winter landscapes where ice turns leafless plants into delicate crystal formations, but they can also be a serious threat to life and limb. The storm that resulted in the scene at right probably began after rain fell into a layer of air near the ground that was only just below 32°F (0°C). The rain may have reached the ground in liquid form, only to freeze soon after, coating all exposed surfaces with a layer of ice. If such conditions persist for a few hours, the amount of ice that accumulates can be so great that it damages outdoor structures as well as creating dangerous conditions for motorists and pedestrians.

↑ **The weight of winter** The weight of the ice buildup in a 1986 ice storm in North Dakota, USA, became so great that the powerlines in the background of this photograph collapsed. Crippling power disruptions are one of the worst effects of this type of weather. The ice of mountain glaciers and in Antarctica can sometimes be so compacted that it becomes solid, turns green and cannot be broken, even with a pickax.

← **No time to be abroad** Heavy snowfalls and strong winds in an urban environment can result in chaotic conditions, with public and private transport disrupted and hazardous conditions created for anyone engaged in outdoor activities. For pedestrians, very windy conditions—the wind-chill factor—exacerbate the physically unpleasant effects of an already low temperature. As well as being physically uncomfortable, intense cold impairs nerve function and reduces manual dexterity.

→ **Wind sculpture** These curious trails in the snow near Spitzbergen on the Arctic Sea island of Svalbard are known as *sastrugi*, from the Russian word for "groove." Despite their even, sculpted appearance they are a natural phenomenon, formed by wind erosion of the snow's surface. Fine snow driven by the wind helps form these sharp irregular ridges and furrows. Blizzards, severe frosts and bitter temperatures are features of the long winter months in the high latitudes of the Russian Arctic. Intense winds can blow for several days at a time.

← **Iced-up marina** Blizzards can create hazardous conditions for vessels of all kinds. Apart from wind damage, a buildup of ice and snow on decks and in riggings may cause them to collapse.

→ **Icy "feathers"** When a high wind picks up tiny droplets of supercooled water and drives them along, the droplets sometimes lodge on solid objects to form these peculiar formations popularly known as ice feathers.

Tornadoes

Tornadoes are commonly associated with a particularly severe type of storm known as a supercell thunderstorm. Such storms are usually characterized by extremely powerful updrafts that sometimes extend to the top of the cloud, producing a bulge in the classic anvil shape, called an overshoot. As the wind speed increases rapidly with height, and the wind changes direction, the updraft near the storm's center rotates rapidly—a phenomenon called a wind shear. This rotational characteristic is one of the main forces behind the savage, spinning energy of the tornado. The power of a tornado is emphasized by the deafening roar that usually accompanies it; the sound of the wind can be heard several miles away and is at its peak when the tornado is touching down to the ground.

THE FUJITA SCALE		
Scale no.	Speeds (mph [km/h])	Damage type
F0	40–73 (64–117)	Light
F1	74–112 (118–180)	Moderate
F2	113–157 (181–251)	Considerable
F3	158–206 (252–330)	Severe
F4	207–260 (331–417)	Devastating
F5	more than 260 (417)	Incredible

PUTTING TORNADOES IN PERSPECTIVE

The scale above classifies the intensity of tornadoes by analyzing the nature of the wreckage trail they leave behind them. When an F5 tornado moves across a populated area, almost total destruction can result, as happened in the devastating Tri-State Tornado of March 1925 in the USA. The scale was developed by Professor Theodore Fujita (1920–98), a meteorologist at the University of Chicago. Measuring the wind speed of tornadoes can, at best, be an estimate, as no meteorological instrument could withstand the devastating force of the maelstrom in a tornado's funnel, as it sweeps up all in its path, carries it upward or onward, and dumps it far from its origins. Most tornado damage results from these high, rotating winds, but some is caused by extreme differences in air pressure.

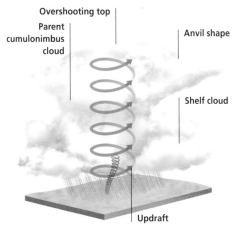

Overshooting top
Parent cumulonimbus cloud
Anvil shape
Shelf cloud
Updraft

← **From thunderstorm to tornado** A supercell thunderstorm is a severe storm that contains a strong rotating updraft called a mesocyclone. Under the right conditions this system extends downward to become more compact, causing it to rotate faster, finally reaching the ground as a tornado. Large hailstones and intense lightning activity sometimes accompany tornadoes, producing even more damage. Two signs are useful in assessing whether a storm will develop into a tornado. The first is the overshoot phenomenon, when the usual flat top of the anvil develops an ominous bulge. This indicates that the rush of air near the center of the storm is so powerful that it has pushed through the troposphere into the stratosphere. The second is the development of mammatus clouds.

← **Watching and waiting** A large tornado, accompanied by intense lightning activity, looms ominously above a small farm. Regular weather forecasts are vitally important when tornadoes occur at night—without them, the only warning of the tornado's approach to a person on the ground may be when it is sighted in the glow of a lightning flash. When meteorological observations indicate the buildup of tornado-like conditions, they are monitored carefully.

↓ **Moving column of water** Waterspouts, such as this one over the James River in Virginia, USA, are normally weaker cousins of the tornadoes that occur over ocean areas and do not usually require a supercell to form. Tornadoes moving over lakes or rivers can draw up water to produce a waterspout.

→ **Formation of a tornado** The distinctively shaped funnel cloud that descends from the base of a severe thunderstorm is what most people associate with a tornado. Very low air pressure inside this funnel causes atmospheric moisture to condense, making the funnel visible. Sometimes debris drawn aloft by the powerful spiral updraft can also color the funnel: the color depends on the kind of dirt and debris that is collected. From Earth, the bulge at the base of the thunderstorm is known as the wall cloud. Tornadoes typically look like an elephant's trunk, hanging from the base of the storm cloud that generated them. Severe weather conditions may produce a line of storms that result in a series of tornadoes. This can be particularly damaging and deadly.

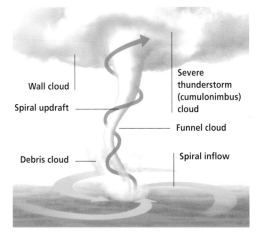

Wall cloud

Spiral updraft

Severe thunderstorm (cumulonimbus) cloud

Funnel cloud

Spiral inflow

Debris cloud

Tornado Alley

Tornadoes have been reported in every continent except Antarctica, but the United States experiences these devastating storms most frequently. The American Meteorological Society defines Tornado Alley as the area that includes the lowlands of Mississippi, the Ohio and the lower Missouri river valleys. Depending on the time of year, the borders of Tornado Alley extend from Iowa and Nebraska in the north to central Texas in the south. As the Great Plains heat up during summer, the air expands and rises, sucking more air in to take its place. Tropical moist air from the Gulf of Mexico blows into the Plains and collides with cold, dry air from the Rockies to the west. A unique combination of moisture supply, and a drying and cooling middle atmosphere provides the ideal conditions for tornadoes. In April 1974, these conditions converged to create the country's largest tornado outbreak. In just over 16 hours, 148 tornadoes hit 11 states, killing 315 people and injuring more than 5300.

↓ **A cloud's mighty power** A large and well-developed tornado over the high country of New Mexico siphons up vast amounts of loose earth and generates an extensive dust cloud beneath the parent thunderstorm. Most tornadoes last only a few minutes and have a path of 165 feet (50 m) wide and 3 miles (5 km) long, so their damage trail is narrow. The most severe, however, can last for an hour and have damage paths 1 mile (1.6 km) across and 60 miles (100 km) long.

↑ **Before and after** Computer-enhanced satellite images of the La Plata area of Maryland indicate that a large area of vegetation—shown as red—has been ripped up and swept away by a fierce tornado. Airborne radar developed by the US military during the Second World War has been adapted to enable more accurate severe weather warnings.

→ **Unstoppable destructive force** A photograph of a tornado sweeping through Pampa, Texas, in 1995 provides an insight into the terrifying experience of being in the path of one of these wild winds. Debris is thrown in all directions, extending well outside the area of the funnel. The color of a tornado's funnel is determined by the kind of dust and debris it has gathered up in its path.

MOST DEVASTATING TORNADOES

Year	Place	Devastation	F Scale
1925	Tri-State (Mo, Ill, In)	695	F5
1840	Natchez, Miss	317	Unknown
1896	St. Louis, Mo	255	F4
1936	Tupelo, Miss	216	F5
1936	Gainesville, Ga	203	F4
1947	Woodward, Ok	181	F5
1980	Amite, LA; Purvis, Miss	143	F4
1899	New Richmond, Wis	117	F5
1953	Flint, Miss	115	F5
1953	Waco, Tx	114	F5

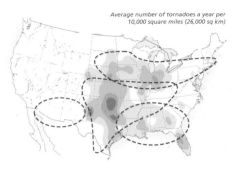

Average number of tornadoes a year per 10,000 square miles (26,000 sq km)

☐ Fewer than one
One

▨ Three
▨ five
▨ Seven

◼ Nine
··· Peak seasons

TORNADOES: WHERE AND WHEN

This map shows the vast area of the United States where tornadoes are most likely to occur. It also indicates the average number that appear annually. Over the southern states, most tornadoes form during January, February or March, but farther north this period gradually slips back to later months.

Aftermath of the deadly Tri-State Griffin, Indiana, was a scene of despair and desolation on March 23, 1925, five days after the great Tri-State Tornado ripped through the town. The tornado wreaked havoc through Missouri, Illinois and Indiana for over three hours, killed 695 and injured more then 2000.

Tornado watch

Situations that may result in the formation of tornadoes are carefully monitored by national weather services. When meteorologists believe that atmospheric conditions are suitable for the generation of a tornado over a particular area, they issue a tornado watch to media outlets. When a weather observer sights a tornado approaching, a tornado warning is issued. While the United States is the most tornado-prone country, experiencing about 750 tornadoes each year, tornadoes also occur in Australia, southern Asia and, occasionally, in other regions such as Europe and the United Kingdom.

→ **Results of the storm's fury** Automobiles were flung into the air and landed amid other debris that resulted from a tornado that passed through Lake Osceola, Florida, USA, in 1998. Properties in the path of a tornado often suffer severe damage, but accurate and timely warnings can save human lives.

↓ **Intrepid chasers capture the moment** A storm chaser is here observing the development of a severe thunderstorm in Kansas, USA. Some storm chasers have produced remarkable photographic and video records of tornadoes.

Identifying tornado potential Meteorologists track severe thunderstorms with the most up-to-date electronic equipment; the output is carefully interpreted and disseminated to the public.

WATCHES AND WARNINGS

A tornado watch is issued when meteorological conditions indicate that tornadoes are possible in a given area. Residents should remain alert for approaching storms and be aware of changing weather conditions. When a tornado has been sighted or indicated by weather radar, a tornado warning indicates imminent danger to life and property in the path of the storm. In this case, residents should move to a predesignated safe place.

Black twister This tornado's spiral updraft has collected a vast amount of debris, contributing to the clearly visible funnel around its base. More accurate predictions of these potentially deadly tornadoes became a reality in the second half of the twentieth century, with the development of Doppler radar and advances in computer technology that allowed meteorologists to create and analyze models of severe weather conditions.

Hurricanes

Extreme rainfall, unbelievably high waves, winds of incredible ferocity —nothing can compare to the destructive potential of huricanes. Known as cyclones in Australia and typhoons in southeast Asia, these intense storms form over the warm waters of tropical oceans. The most favorable conditions for their development are found between 5 and 15 degrees of latitude, slightly away from the equator where the Coriolis force is strong enough to help spin up the hurricanes and sea temperatures are above 79°F (26°C). Once formed, they may last for days or even weeks before they sweep poleward or cross over the land with potentially disastrous results. Around 80 hurricanes form each year, with around 35 forming near southeast or southern Asia, 25 near the Americas and the remainder shared across the southern Indian and Pacific oceans.

The map at right shows the main areas of hurricane distribution and the major storm paths. The hurricane season runs from June to November in the northern hemisphere and from November to May in the southern hemisphere.

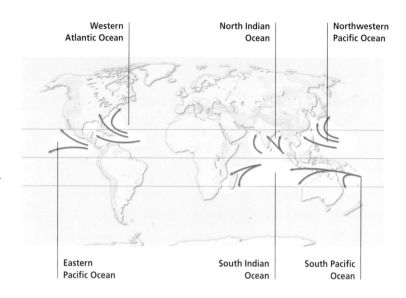

Western Atlantic Ocean   North Indian Ocean   Northwestern Pacific Ocean

Eastern Pacific Ocean   South Indian Ocean   South Pacific Ocean

THE SAFFIR–SIMPSON SCALE

In the American region the intensity of a hurricane, as determined by its maximum sustained wind speed, is given by a five-point scale. Rare Category 5 storms, such as Hurricane Camille that struck the Gulf coast and Virginias with catastrophic effect in 1969, are the most severe.

↓ **Hurricane Lili** There is an eerie beauty in satellite photographs that capture the awesome power of hurricanes. This view of Hurricane Lili, shown before it swept across Louisiana, USA, was captured from the Space Station.

THE SAFFIR–SIMPSON SCALE

Pressure (hectopascals)	Wind speed (mph [km/h])	Storm surge (ft [m])	Damage type
1 more than 980	74–95 (118 –152)	4–5 (1.2–1.6)	Minimal
2 965–110	96–110 (153 –176)	6–8 (1.7–1.6)	Moderate
3 945–964	113–130 (177–208)	9–12 (2.6–3.7)	Exstensive
4 920–944	131–155 (209–248)	13–18 (3.8–5.4)	Extreme
5 less than 920	more than 155 (248)	more than 18 (5.4)	Catastrophic

INSIDE A HURRICANE

Long-lived clusters of towering thunderstorms draw their energy from warm tropical oceans. The air spirals (counter-clockwise in the northern hemisphere and clockwise in the southern hemisphere) into the center, or eye, of the developing hurricane, accelerating as it does so. The air then spirals upward in the torrential rain, producing eyewall cloud that surrounds the tranquil eye of the storm. In the upper levels the air spirals outward away from the hurricane's center.

↑ **Hurricane over Madagascar** This satellite view of Tropical Cyclone Dina was taken in January 2002, soon after the system had started weakening, although it was still producing surface winds of 130 miles per hour (210 km/h). The beauty of the spiral disguises the enormous power of the storm. An average hurricane is about the length of Scotland; its "eye" can be up to 30 miles (50 km) in diameter.

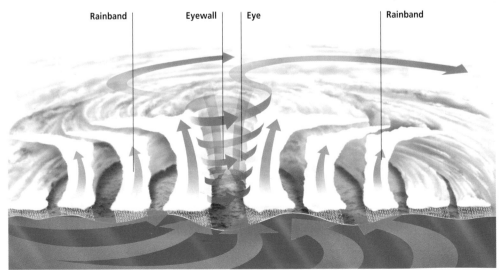

Rainband   Eyewall   Eye   Rainband

Hurricane impact

Hurricanes are among the most destructive forces in nature. Winds from these powerful systems can gust up to 190 miles per hour (300 km/h). While tornadoes can create even stronger winds, they rarely last for more than a few hours; by contrast, hurricanes can last for weeks. The destructive effect of hurricanes has been chronicled for thousands of years: fleets have been sunk and cities destroyed by vicious winds and mountainous seas. The Japanese term *kami-kazi*, meaning "divine winds," referred to the hurricane-like storms that sank the fleets of the Mongol emperor Kublai Khan in 1274 and 1281 as they prepared to attack Japan. Damage caused by hurricanes ranges from complete devastation to relatively minor inconvenience. The greatest destruction is usually from water damage resulting from a storm surge that causes a rise in ocean level as the hurricane approaches a coastal feature, bay or island. The height of, and potential damage from, the storm surge depends upon the slope of the ocean floor along the coast. As hurricanes move inland from the coast, they become low-pressure systems or rain depressions. These often bring heavy rainfall to inland areas and cause widespread flooding.

↓ **Fury of the hurricane's fringe** The mighty waves that hurricanes spawn can wreak havoc for hundreds of miles in every direction. Hurricane Iris bypassed Jamaica on its way toward the Yucatan Peninsula, Mexico, in October 2001, but the force of its waves was enough to drive this cargo ship aground in Kingston Harbor.

→ **Dumped ashore** The fury of a hurricane can force the sea several miles inland—a fact clearly demonstrated by this image of a yacht that was swept across a section of the Florida Everglades during Hurricane Andrew in 1992. Hurricane Andrew was the costliest in United States history, with a damage bill of 30 billion US dollars .

→ **Forest flattened** Entire forests, such as this one in Kent, were devastated in October 1987 when the most intense storm in nearly 300 years struck southern England.

↓ **Extra-tropical** With the right conditions, intense low-pressure systems outside the tropics can deepen to become as dangerous as midlevel tropical hurricanes. This boat was one of 90 craft damaged when an extra-tropical storm hit Padanaram Harbor, Massachusetts, USA. Pleasure craft are particularly vulnerable to damage from the winds and high seas that come with intense lows.

Hurricane watch

Specialized hurricane, typhoon and tropical cyclone warning centers in national weather services around the globe are continuously on the watch for the first signs of developing tropical storms. Every available technology is utilized—ranging from satellites orbiting high above the atmosphere, to specially strengthened hurricane surveillance aircraft, to weather-watch radar, weather-observing ships, drifting buoys and automatic weather stations on remote islands—to make sure no hurricane goes undetected. Once identified, advisories and warnings are regularly and meticulously prepared, then rapidly dispatched around the world. Using a wide range of communication techniques, these messages are sent to all threatened communities and ships at sea.

The internet is an increasingly popular source of hurricane warnings, but radio and television broadcasts are still the public's preferred source of information. Recorded telephone and SMS services are also in widespread use. Ships at sea rely upon fax and radio broadcasts, often delivered via satellite. In some countries, the impending landfall of a hurricane is signaled by strategically placed sirens along the coastline. Once warnings have been issued, local emergency service groups swing into action and communities begin preparing for the storm. This can include boarding-up windows, tying-down outdoor equipment or, in severe cases, mass evacuation of the local population.

SIGNS FROM THE SEA
While sophisticated technology is of paramount importance in identifying hurricanes, simpler signs include swells at sea. A hurricane at sea produces a swell that spreads from the center of the system and often runs well ahead of the storm. If a large swell is observed, a hurricane may be approaching.

August 25

August 24

→ **Caribbean disaster**
The Caribbean islands are especially vulnerable to hurricanes as they sweep westward across the Atlantic. This debris is the aftermath of a hurricane that hit Saint-Martin in Anguilla.

→→ **Tracy's aftermath**
Cyclone Tracy hit Darwin, on Australia's northern coast, on Christmas Day 1974, killing 49 people. Although relatively small, its destructiveness resulted from a direct hit on the town.

August 23

TRACKING A HURRICANE

The destructive path followed by Hurricane Andrew toward Florida, USA, is shown in this composite image which spans a 48-hour period of its lifetime. Hurricane Andrew began as a tropical storm that emerged from an atmospheric disturbance over the North Atlantic Ocean on August 17, 1992. It reached hurricane strength on August 22.

Within two days the hurricane smashed into the coast of the United States, making it only the third Category 5 hurricane on record to cross the country's mainland. The hurricane passed over Florida on a path toward the Gulf coastline where it wreaked further havoc. The United States dollar 30 billion damage bill for this devastating event would doubtless have been far greater but for the warnings issued well beforehand. Satellite images such as these are updated regularly when a hurricane is identified and tracked.

August 23 Andrew continued to strengthen as its eye crossed over the northern end of the island of Eleuthera in the Bahamas.

August 24 Shortly after reaching peak intensity, Andrew smashed into the coast of Florida with sustained winds of 165 miles per hour (265 km/h).

August 25 Despite weakening slightly as it headed toward the Gulf coast, Andrew caused considerable damage in Louisana.

→ **Wholesale destruction** In only a few minutes Hurricane Andrew turned this Miami trailer park into a chaotic scene of devastation. Only the sturdiest buildings can withstand the forces of high-category hurricanes.

→→ **Stormy Daniel** This infrared image from July 31, 2000 captures the intensity of Hurricane Daniel, which traveled over 2000 miles (3000 km) from the southwest to threaten Hawaii.

Floods, landslides and avalanches

Heavy rain and snow can leave many problems in their wake, ranging from floods to landslides and—in the case of snow—avalanches. Floods alone account for 40 percent of casualties from natural disasters. Communities most at risk are those nestled in valleys against the sides of steep terrain; settlements that lie in broad river valleys and river deltas are also vulnerable to flooding.

In some areas of the world, floods are part of the natural weather cycle. In the Nile valley, for example, regular flooding sustained agriculture for thousands of years. Today, many tropical regions depend on the floods that follow monsoonal rains to nourish crops. Less predictable, and potentially more damaging, are the flash floods that result when intense, short-term rainfall cannot be dispersed by soil absorption, runoff or drainage. Broadscale flooding, by contrast, is normally associated with a frontal system such as a cold front or low-pressure cell that produces prolonged rain over an extensive area. It may take several weeks to reach its peak.

→ **Riverside risk** The Ecouen district in the Paris region of France was inundated by murky floodwaters in June 1992. Many of the world's largest settlements have been built on riverbanks and when a major flood hits, thousands can suddenly be left homeless when the banks break.

↓ **A frightening warning** Although snow may not appear to be a weighty substance, when it is compacted and accumulated to several feet in depth, it becomes extremely heavy. Avalanches pick up rocks and trees as they sweep down steep slopes, increasing their destructive power as they go. This scene of devastation followed an avalanche at Montroc in the French Alps in 1999. A massive air blast sometimes travels ahead of a snow avalanche—the so-called sigh of the avalanche.

FORMATION OF A FLASH FLOOD

The most common cause of flash flooding is a slow-moving thunderstorm that deposits huge amounts of rain over a small area in a short period. When humid air is blown toward a mountain, it rises and may develop into a storm. If winds keep the storm stationary, torrents of rain rush down the mountainside into the valley below. Surprisingly, slow-moving storms can also cause flash flooding in deserts. Dry, baked soil absorbs little rain, and a downpour can quickly turn dry riverbeds into raging torrents. More people drown in North American deserts than die of thirst. For similar reasons, flash flooding is increasingly common in cities. As more ground is leveled off and covered with asphalt and concrete, less rain can be absorbed. Drains overflow and water can rapidly flood along the streets.

WHEN THE EARTH MOVES

There are two broad types of landslide. The first are those where rainfall weakens a steep section of a hillside to a point where the soil and vegetation break free and slide down a slope. The second and equally devastating type forms when heavy rainfall soaks into the ground and turns it into slurry. This forms a river of mud and rocks which pours down a hillside, burying everything in its path. In May 1998, for example, after 24 hours of torrential rain, a landslide near Naples in Italy killed 20 people and destroyed hundreds of homes.

City of rubble This landslide in Vargas State, Venezuela, caused extensive damage to a sizable city in December 1999. Very heavy rainfall can cause devastating landslides that strike with little warning.

INGREDIENTS FOR A SLAB AVALANCHE

A varied combination of events can culminate in an avalanche above the snowline. Initially a good depth of snow is required to establish a substantial base. Heavy follow-up snowfalls can lead to huge accumulations of snow on the upper reaches of the mountain slopes. If there is heavy rain over the lower slopes, the support for the large snow mass farther upslope is weakened. Sometimes, if the slope is steep enough and the snow deep enough, an avalanche can occur even without the downslope rain. The unsupported mass of snow can spontaneously break free and come cascading down the mountainside. This may be triggered by the smallest of vibrations: a loud noise or the sound of a skier can be sufficient. A strong wind gust or a temperature rise are other common triggers.

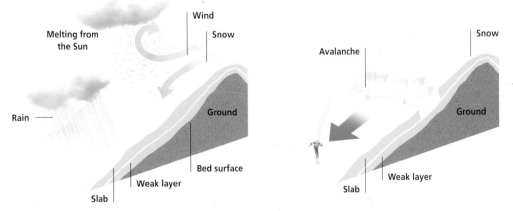

Droughts

Although not always as dramatic or headline-making as other natural disasters, droughts cause billions of dollars worth of economic losses in the countries where they occur. In poor nations droughts lead to widespread famine with malnutrition-related illnesses lasting long afterward. In the long term, the environment suffers badly: the ground is denuded by starving stock, leading to widespread erosion and a decrease in the animals' fertility. Drought may cause native animals to lose their habitat and food supplies, and thus increase the rate of extinction. This cycle of degradation leads to poor crop and native grass regrowth in ensuing years. The plants are less able to cope with the next dry spell, leading to worse droughts and further famine in the future. Moreover, parched topsoil and tinder-dry vegetation create perfect conditions for dust storms and wildfires.

DROUGHT: NOWHERE IS SAFE

Droughts are most likely to affect countries in the middle latitudes—the semi-arid and Mediterranean climate zones which receive much of their rain from transient weather systems. However, few places are safe from drought. Even the Amazon jungle and the tropical islands of Indonesia have experienced serious droughts in recent years. Severe droughts last for many years.

↑ **In search of water** Village women carry earthenware pots in search of water in drought-stricken Sindh province, Pakistan, in May 2000. Drought is all too frequent in the region.

← **Handfeeding** The crippling drought that struck many parts of Australia in 2002 meant these sheep in western New South Wales had to be handfed.

AT THE MERCY OF EL NIÑO

The El Niño phenomenon, which occurs when a large expanse of warm water accumulates off the Peruvian coast, sets off a chain reaction that brings drought to many countries, including Australia.

THE CYCLE OF DROUGHT IN AUSTRALIA	
Years	**Region affected**
1888	All states except Western Australia
1895–1903	All states
1911–1916	Most states
1918–1920	All states except Western Australia
1939–1945	All states
1958–1968	All states
1982–1983	All of eastern Australia
2001–2003	All states

WHEN FARMLAND TURN TO DUST

Drought is not merely low rainfall. Rain is unevenly distributed throughout the world and some areas will always receive less than others. Drought is a sustained and abnormal deficit of rain, based on the expected rainfall of an area at a given time of year. In the United States, the term is used when an area receives 30 percent or less of its normal rainfall over a minimum 21-day period. In India, by contrast, drought is declared if annual rainfall is less than 75 percent of the aveage.

← **Waiting for the monsoon** An Indian woman carries a pot of drinking water as she walks along the dried-up Osman Sagar lake in the southern Indian state of Andhra Pradesh, Hyderabad, in June 2003. Excessive heat and dry conditions in areas such as this can be broken by the monsoon.

↓ **Starved of water** As a drought takes hold, near-surface groundwater is lost. Plants become stressed and leaves starved of water lose their chlorophyll.

Heatwaves and wildfires

Prolonged heatwaves cause many deaths each year, particularly amongst the elderly, infirm and very young. In France in 2003, more than 3000 people were reported to have died as a result of an extended heatwave. Because of global climate warming, the incidence of heatwaves is expected to increase in coming decades.

Hand-in-hand with heatwaves come wildfires. These occur most frequently and to most devastating effect in California, USA, southern and eastern France and large parts of Australia. The climates of these regions feature winter rainfall and summer drought, leading to an annual buildup of highly flammable vegetation. When strong winds and high temperatures coincide, a wildfire outbreak is likely. The intense heat generates strong local thermals which send burning embers far up into the sky, igniting new fires well ahead of the main fires. In addition, fire-induced localized wind patterns, in combination with local topography, may generate rotating "fire tornadoes" that spin ahead of the main firefront.

SUMMER IN THE CITY

Heatwaves in urban environments produce a number of familiar summertime phenomena. People spend more time outdoors seeking relief from the heat—often near large or small bodies of water. Electricity networks are prone to fail from the extra load imposed by millions of air conditioners. And road congestion increases as automobiles overheat and break down.

↓ **Practical Parisians** A sweltering heatwave that hit Paris in July 1995 sent the city's residents seeking relief wherever they could find it. Public fountains had a cooling effect on people of all ages.

THE CURSE OF THE FIREBUG

In the natural world, almost all fires are caused by lightning strikes. In fact, many ecosystems need periodic fires to regenerate certain types of vegetation. In modern times, lightning still starts fires but unfortunately the majority of blazes are deliberately lit, many during times when existing fires are out of the control of teams of firefighters. Other human activity that has increased the frequency of wildfires includes deliberate burnoffs that get out of control, unextinguished campfires and discarded cigarette butts. If small blazes such as these are not controlled quickly, they can become raging infernos.

↑ **Watching the inferno** Fire threatens homes in the Simi Valley, near Yosemite National Park, California, USA, in October 2003. Multiple wildfires raged through much of California during this outbreak, affecting more than 250,000 acres (100,000 ha) of land and destroying property.

← **A city threatened** These fires destroyed forests near the heart of Canberra, the Australian capital, in 1985. Even worse fires returned in 2003, claiming the lives of at least four people, injuring more than 150 and razing hundreds of homes.

→ **Raging inferno** Fierce flames completely engulf an entire forest during the Shoshone Fire of July 1988 in Yellowstone National Park, Wyoming, USA. Many sections of the park were closed during this fire. Large wildfires can quickly race out of control and destroy vast tracts of plantation or native forest in the space of only a day or two.

Dust storms

Mention dust storms and people instinctively think of the great sand deserts of the world. And it is in these hot, dry and sparsely vegetated regions that dust storms are most frequent and pervasive. Many of these blinding storms are triggered by strong, seasonal, local winds that have been given names: among them, the haboob of northern and eastern Africa, and the shamal of the Persian Gulf region.

Dust storms are also irregular but fierce visitors to other parts of the world. Given the right combination of prolonged dry weather, high temperatures and strong winds, dust storms can become the scourge of the countryside. During the "dust bowl" years of the 1930s in the USA so much dust was picked up by the severe storms that it was swept all the way across the Atlantic Ocean to Europe. And in Australia, severe dust storms that occurred over the southeast during the prolonged drought of 1983 carried vast clouds of dust across the Tasman Sea to produce dust-colored "red snow" on the glaciers of the South Island of New Zealand.

THE DRIFTING DUST

In many cities within or near desert regions there is a continuous rain of fine dust from the surrounding countryside, making it virtually impossible to dry clothes outdoors or to keep automobiles clean. Jet aircraft that operate in these regions require servicing twice as frequently as in more benign climates because the all-pervasive dust wears away the edges of their toughened metal fan blades. Some large dust storms are so extensive that they are visible from space. In these cases, spectacular satellite photographs provide valuable information about the precise dimensions and movements of the dust clouds.

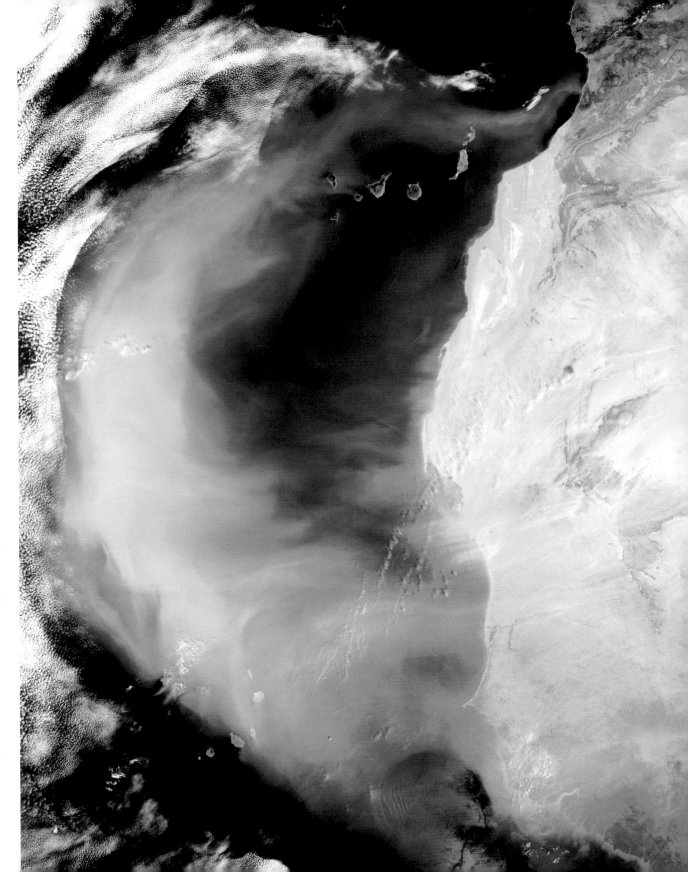

Storm season In desert regions sand storms form in the warmest months, particularly once the winds freshen. The people of Djanet in Algeria prepare for this regular event. Sand can travel several thousand miles and remain airborne for days when it is whipped up by the local sirocco winds

Beijing traffic hazard A blinding sand storm that struck the Chinese capital of Beijing in 2001 turned daytime into an eerie red twilight as it blotted out the Sun. High pollution levels intensify the strange light.

Terrible days of the "dust bowl" Huge dust clouds cast an ominous pall over houses in rural Springfield, Colorado, USA, in 1935. These dust storms were locally known as " black blizzards."

Dust cloud traveling on the wind A satellite photograph shows a dense pall of dust that originated over the Sahara Desert and enveloped the Canary Islands in January 2002. Thick plumes of dust stream across the Atlantic Ocean.

Record-breaking weather

Tall tales and true have been told about the weather for as long as there have been people around to tell them. Sorting the "tall" from the "true" is not easy—few records have been confirmed by reliable meteorological measurements. Systematic observations started only in 1814 when the Radcliffe Observatory in Oxford, England, began recording changes in weather. In the United States, daily records started in 1885 in an observatory founded by Abbott Lawrence Rotch in Milton, Massachusetts. This observatory—the Blue Hill—continues to keep meteorological records and itself holds the record for the longest continuously operating weather-observing station at the same location in the United States. Extremes of weather can be officially cited as records only if the weather station that recorded them has a long-term set of weather measurements. Just how long weather stations should maintain data before declaring records remains a matter of debate, but the consensus is that at least 10 years' measurements are required before an extreme reading is declared to be a record. The accurate records maintained by today's weather stations across the world enable comparisons to be made and extremes of weather to be documented. Together, these extremes reveal the enormous power of the forces that contribute to our weather.

A WORLD OF EXTREMES

The highest air temperature ever recorded—admittedly questioned by some—was 136°F (57.8°C) at Al Aziziyah in Libya on September 13, 1922. At the opposite end of the scale is Vostok Base in Antarctica which has an annual average temperature of –72°F (–58°C). On July 21, 1983 it recorded the world's lowest temperature, –128.6°F (–89.2°C). The most extreme temperature range occurs at Verkhoyansk in Siberia—down to –90°F (–68°C) in winter and up to 98°F (37°C) in summer. The record for the number of tornadoes is held by the northeast region of Colorado in the Great Plains of the United States with an average of 1.5 tornado days per year. Wind speeds of up to 310 miles per hour (500 km/h) have been recorded.

Strong winds are not confined to tornadoes. The highest non-tornado wind gust of 231 miles per hour (371 km/h) was recorded at Mount Washington, USA, on April 12, 1934. Stories of large hailstones are like tales of fish that got away: those that are collected and measured are always smaller than those that escaped. The largest authenticated hailstones fell in the Gopalganj district in Bangladesh on April 14, 1986 and killed 92 people. The record for the largest number of days of rain per year goes to Mount Waialeale, on the island of Kauai in Hawaii, where rain falls on average 350 days each year. By contrast, the Atacama Desert in Chile is the world's driest place, with effectively no rainfall at all.

Greatest temperature change in one day 100° F (55.6° C), a temperature drop from 44° F (6.7° C) to –56° F (–49° C), on January 23–24, 1916 in Browning, Montana, USA

Most snow on ground 451 inches (11,455 mm) in March 1911 at Tamarack, California, USA

Highest winds in a landfalling tropical system 200 mph (322 km/h) wind with gusts up to 210 mph (338 km/h), on August 17–18, 1969 along the Alabama and Mississippi coasts, USA, during Hurricane Camille

Driest location The Atacama Desert in Chile has virtually no rainfall (0.003 inches [0.08 mm] annually), except for a passing shower several times a century

Tropical rains Java, Indonesia, epitomizes the tropical monsoonal regions where rainfall is highest.

Siberian freeze The open, flat lands of central Siberia are exposed to some of the coldest temperatures on Earth.

Heat of the dunes The world's highest temperature was recorded at Al Aziziyah, in the Libyan Sahara.

Rarely dry Water cascades down the slopes of Mount Waialeale, Hawaii—the wettest place on Earth.

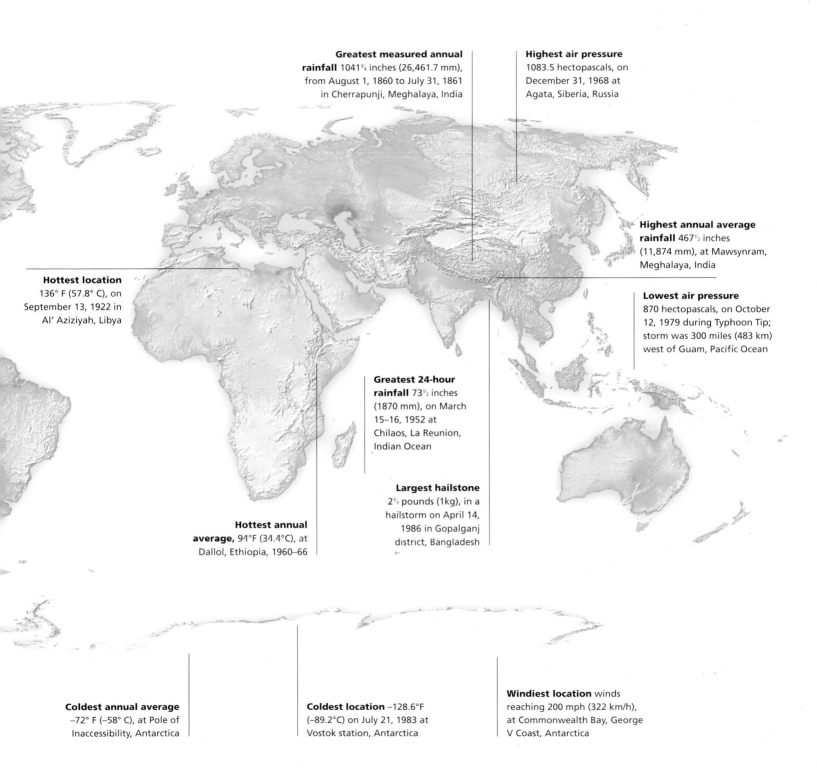

Greatest measured annual rainfall 1041¾ inches (26,461.7 mm), from August 1, 1860 to July 31, 1861 in Cherrapunji, Meghalaya, India

Highest air pressure 1083.5 hectopascals, on December 31, 1968 at Agata, Siberia, Russia

Highest annual average rainfall 467½ inches (11,874 mm), at Mawsynram, Meghalaya, India

Hottest location 136° F (57.8° C), on September 13, 1922 in Al' Aziziyah, Libya

Lowest air pressure 870 hectopascals, on October 12, 1979 during Typhoon Tip; storm was 300 miles (483 km) west of Guam, Pacific Ocean

Greatest 24-hour rainfall 73½ inches (1870 mm), on March 15–16, 1952 at Chilaos, La Reunion, Indian Ocean

Largest hailstone 2¼ pounds (1kg), in a hailstorm on April 14, 1986 in Gopalganj district, Bangladesh

Hottest annual average, 94°F (34.4°C), at Dallol, Ethiopia, 1960–66

Coldest annual average −72° F (−58° C), at Pole of Inaccessibility, Antarctica

Coldest location −128.6°F (−89.2°C) on July 21, 1983 at Vostok station, Antarctica

Windiest location winds reaching 200 mph (322 km/h), at Commonwealth Bay, George V Coast, Antarctica

Extreme weather: USA

As the "scorekeeper" of extreme weather events in the United States, the National Climatic Data Center tracks and evaluates climatic events that have great economic and social impact. Between 1980 and 2003, the United States sustained 57 weather-related disasters in which overall costs and damage exceeded one billion dollars. Of these, 14 were hurricanes, most of which hit the coast in the Florida–North Carolina region. More than 3500 people lost their lives to these hurricanes, the most deadly being Hurricane Hugo in 1989, which claimed 86 lives. Five tornado outbreaks were recorded, the most deadly of which was a series of tornadoes and associated flooding in Mississippi, Ohio, and adjacent states in 1997 that resulted in 67 deaths. At the other extreme, drought and fire ravaged a number of states: in 2002, drought over large parts of 30 states cost an estimated 10 billion dollars in damage; drought and heatwaves in 2002 in the southern states caused significant losses to agriculture and claimed around 140 lives.

→ **Ice mountain** A severe blizzard in January 1953 left the observatory buildings on Mount Washington, New Hampshire, coated in a thick crust of ice. Temperatures here have fallen to -46.5°F (-44°C).

↓ **A mighty flood** Rapidly rising floodwaters swept a train off this bridge over the Arkansas River at Puelo, Colorado, in June 1921. The force of the flood left trees and other debris strewn across the wrecked, steel-framed bridge.

WILD WATERS OUT OF THE NIGHT
A feature of many extreme weather events is the frightening speed with which they occur. On the afternoon of July 31, 1976 there was no indication that anything was out of the ordinary in the Big Thompson River Canyon, Colorado. But between 6.30pm and 10.30pm severe thunderstorms in the catchments of the river dropped a massive 12 inches (305 mm) of rain. The resulting floodwaters roared down the river, breaking its banks and killing 135 of the 2000 people in its path.

THE GREAT WHITE HURRICANE OF 1888

For an intermittent 36 hours, between March 11 and 14, 1888, a massive blizzard engulfed much of the east coast of the USA, producing widespread devastation across the region. The storm, known colloquially as "The Great White Hurricane," dumped an estimated 50 inches (125 mm) of snow in Connecticut and Massachusetts, and 40 inches (100 mm) across New York and New Jersey. Gale-force winds raged, creating huge snowdrifts that paralyzed transport. Food and heating shortages followed, and 400 people lost their lives.

↑ **Wisconsin: the scorched earth** Drought descended on Wisconsin in 1871. Hot southerly winds that sprang up during early October helped fan the flames from a number of small fires into a massive conflagration that swept through the tinder-dry countryside. The blaze destroyed a 60-mile (96.6 km) stretch of land, killing more than 1200 people. The Peshtigo River is said to have boiled in some places, killing those who sought refuge in the water.

← **Urban paralysis** New York's streets were impassable during the 1888 blizzard; the transportation crisis led, eventually, to the construction of the city's subway 12 years later.

Extreme weather: Europe

In Europe, floods are the most common natural disaster. Between 1975 and 2001, 238 major floods were recorded in the region. Since 1990, about 2000 people have died during floods and some 400,000 have been left homeless; in 2002, 15 major floods killed about 250 and affected at least a million others. In Poland, floods in 1997 spread over 2300 square miles (6000 sq km) of land and forced the evacuation of 160,000 people. At the other extreme is an unprecedented increase in warming in the recent past. In the summer of 2003, intense heatwaves struck large areas of western Europe. At least 3000 heat-related deaths were recorded in France alone in just two weeks in August, with infants and the elderly most vulnerable. Although long-term climate change has affected Europe for millions of years, recent events suggest that extreme weather events, with their attendant human and economic costs, may become more frequent and intense. International and regional organizations have set up task forces to monitor and respond to these changes.

↓ **A future of long, hot summers** Parisians seek relief from the heatwave of 2003. The changed temperature extremes imposed on many regions by global warming will be accompanied by a number of unwelcome effects. Heatwaves will become more frequent, more intense and of longer duration, particularly in the midlatitudes. There may be an increase in wildfire outbreaks and droughts; semiarid zones can expect bigger and more frequent dust storms. In cities, hospital admissions due to heat stress can be expected to rise.

↑ **Storm in England** In October 1987, a severe storm battered the south of England. At least 18 were killed and 15 million trees destroyed by winds that reached 110 miles per hour (177 km/h). The last storm of similar magnitude was in 1703.

→ **Frozen Thames** Climate change is not a new phenomenon. This frost fair was held on the frozen River Thames in London during the Great Frost of 1739–40. The Thames froze frequently during the Little Ice Age.

Surviving extreme weather

Humans evolved in the hot, dry savannas of Africa and today find it easier to acclimatize to heat than to extremes of cold. We keep cool by sweating: skin temperature is lowered as perspiration evaporates. In extreme cold weather, heat loss from the body can quickly lead to hypothermia—a reduced, and potentially fatal, body temperature that progressively causes mental and physical collapse. Humans have found ways to live in inhospitable climates and have developed measures to survive extremes of weather. Accurate and timely forecasting of forthcoming storms, tornadoes and hurricanes, floods and wildfires is essential to community preparedness and response. Media reports are critical in disseminating information. Increasingly the internet is an up-to-the-minute source of regional and local weather conditions.

Extreme weather can take many forms, and each requires a specific response. Understanding the phenomenon and its likely ramifications is all-important. Planning is paramount. Each form of extreme weather has its own imperatives. In an earthquake-prone area, an emergency pack with battery-powered torch, water and first-aid supplies should be standard. If caught outdoors in a lightning storm, try to get inside a vehicle or building, stay away from windows and keep down. The best protection in a tornado is to seek shelter: stay inside with the doors closed, as far away as possible from exterior walls and windows. Crouch under furniture and protect your head with pillows or cushions. Do not try to outrun a tornado by driving; try to find a solid structure in which to shelter. If caught in the path of a wildfire, the best protection is to stay in your car.

↓ **Taking refuge** As Hurricane Floyd ravaged the east coast of the United States, this boy and his dog found shelter in the Charleston Coliseum.

↘ **Away from it all** Children from a coastal village in Guatemala sleep while taking shelter from Hurricane Iris in October 2001.

→ **Road becomes river** A motorboat became the preferred form of transport when the Missouri River broke its banks and flooded highways in St. Charles, Missouri, USA. Broadscale flooding such as this can result when rivers overflow from extended periods of heavy rain or as the destructive, long-term aftermath of a hurricane.

Caught in the flood Drivers are at risk from both broadscale and flash floods. Although in many parts of the world they are part of the natural weather cycle, floods account for 40 percent of all deaths from natural disasters worldwide. National weather services can monitor rainfall and advise of potential floods. However, floods remain difficult to predict and prepare for. While levees can be built along riverbanks, they sometimes fail under pressure.

THE HUMAN RESPONSE

Humans' response to weather is not always confined to the extremes. Many common ailments are affected by weather. Arthritis sufferers, for example, often experience more pain in winter, partly because muscles contract over arthritic joints in the cold weather. When the long, dark nights of winter arrive in high-latitude countries and day-length is shortened, depression and suicide increase. This phenomenon is known as seasonal affective disorder, or SAD. Conversely, in the tropics, the period between the dry and wet seasons, when humidity is at its height, has been noted as a time of depression. The arrival of the monsoon generally provides a lift in spirit. In cities, heatwaves seem to trigger violence: the murder rate in New York, for example, jumped 75 percent during the 1988 heatwave. Windy days are often blamed for the misbehavior of schoolchildren, and intense winds such as the sirocco have been considered mitigating circumstances in criminal trials.

→ **At the mercy of the waves** Coastal property is at risk when high winds associated with hurricanes or other extreme storms whip up massive waves such as these. Hurricanes pose the greatest risk to human life when they reach the coastline. Evacuation can minimize loss of life but destruction of property is inevitable in these conditions. Flooding from storm-generated rain adds further damage to property along the coast.

Previous page A penguin observes the weather conditions from its vantage point in Antarctica.

Watching the weather

Weather forecasting is today a complex and highly technical process, relying on observation, satellite photography, radar and computer simulations. It has come a long way since the days of superstition and sky gods.

Weather lore

Ancient peoples often reacted to the weather in a fearful, superstitious manner. They believed that mythological gods controlled elements such as the winds, rain and Sun that governed their existence. When weather conditions were favorable, there would be plenty of game to hunt and fish to catch, and crops would yield bountiful harvests. But their livelihood was at the mercy of wild weather: fierce storms could damage villages of flimsy huts, destroy crops and generate floodwaters that could sweep away stock. In times of drought, food shortages and famine were a constant threat as crops failed and game animals became scarce when their food supplies dried up.

The ancients believed that their weather fortunes were inextricably linked with the moods and actions of their gods. For this reason, they spent a great deal of time and effort appeasing mythological weather gods. The Egyptians celebrated Ra, the Sun God. Thor was the Norse god of thunder and lightning, a god to please so that calm waters would grace their seafaring expeditions. The Greeks had many weather gods, with Zeus the most powerful.

EARLY OBSERVERS

Several early civilizations used astronomical observations to help them monitor seasonal changes in the weather. Chinese observers had, by 300 BC, developed a calendar that divided the year into 24 "festivals" and described the weather associated with each one. The first reference to a rain gauge was in India around 300 BC. Studying the heavens, too, enabled early observers to forecast weather changes: the Assyrians linked haloes with forthcoming rain.

RITUALS TO BRING THE RAINS

Many ancient societies tried to remain on favorable terms with their deities through a mixture of prayer, rituals, dances and sometimes sacrifices. Native North Americans, among others, performed rain dances during times of drought. Some other cultures, such as the Aztecs of Central America, went so far as to offer up human sacrifices to appease their rain-god Tláloc.

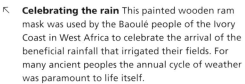

↖ **Celebrating the rain** This painted wooden ram mask was used by the Baoulé people of the Ivory Coast in West Africa to celebrate the arrival of the beneficial rainfall that irrigated their fields. For many ancient peoples the annual cycle of weather was paramount to life itself.

← **Quetzalcoatl, the all-powerful deity** In ancient Aztec society the deity Quetzalcoatl, whose name means "Precious Feathered Serpent," played a pivotal role; he was the creator of life and controlled the rain-bearing winds.

↑ **Gods of the Nile** A thirteenth-century BC mural from the walls of the Sennedjem tomb in Dehr al-Madinah in Egypt shows Re-Harakhty, a blend of the all-powerful Sun god, Ra, and Horus, the god of sky, light and goodness, on a boat on the River Nile. In front is the sacred ibis, representing all that was good in Egypt.

← **The serpent that brings the rains** The importance of water to the survival of the Aboriginal people of Arnhem Land, in the far north of Australia, is celebrated in this rock painting, honoring the Rainbow Serpent. The serpent, said to inhabit the region's permanent waterholes, is believed to send the rains that bring an end to the annual dry season.

→ **Calendar of the Sun** The Sun was central to life in many ancient societies. In this Aztec calendar, carved from stone, the dominant influence of the Sun is shown by its place in the center; the seasons, each reflecting changes in the daily life of the Aztecs, are shown on the perimeter, revolving around the Sun.

↑ **A biblical tale of extreme weather** Noah takes all the world's creatures, two by two, onto his ark in order to ensure their survival of the greatest flood the world has ever seen. According to the Bible, the 40 days and nights of rain were to cleanse the world of its sins. A similar account occurs in Babylonian legend. It is likely that both stories were based on the same historical event: excavations in Iraq have provided evidence of a great flood between 3000 and 2000 BC.

Scientific beginnings

The first scientific attempts to understand weather date back to *Meteorologica* by the Greek scholar Aristotle (384–322 BC). The treatise was an ambitious attempt to describe the physical world, and its title gave rise to the term meteorology. Aristotle's pupil Theophrastus (*c* 372–287 BC) continued his work with *On Weather Signs*, which listed 50 signs of storms, 80 of rain and 45 of wind. Like Aristotle, his observations were mixed, with some shrewd deductions and some misguided premises. Roman scholars, too, showed interest in meteorology. Pliny the Elder's (AD 23–79) monumental work *Historia Naturalis* drew together records, observations and superstitions from Egypt and Babylon, Greece and Rome—some accurate, others perpetuating the myths of earlier times. When the Roman Empire collapsed in the fifth century AD, scientific endeavor was confined to the Islamic world.

THE WEATHER AND RELIGION
The forces of nature have shaped the lives of people from the earliest times. Indeed, the first "meteorologists" were the priests and shamans of ancient communities. All religions recognize the power of the elements and most scriptures contain tales about, or prophecies foretelling, great natural disasters sometimes visited upon a community because of the sins of its citizens.

GIVING THE WIND A FACE

A handsome example of an early civilization's attempt to explain the weather by associating a particular phenomenon with its own mythical deity is the structure known as the Tower of the Winds. Built of marble in around 100 BC, it still stands in a well-preserved state in Athens, Greece. On each side of the octagonal building a male personification of the wind from the direction the side faces is depicted in a decorative frieze. In its original state, there were sundials on the external walls and a water clock inside. Centuries of earth and rubble were cleared from around the tower in 1840 and it was restored to its present condition during the First World War and again during the mid-1970s.

↑ **Superstition rampant** Failing a rational explanation for harsh weather, communities in medieval Europe often burnt at the stake women believed to be witches so that they could no longer use spells to affect the weather. The Middle Ages stifled scientific enquiry, including rational debate about weather phenomena.

← **The Khan's fleet** In 1896, the artist William Henry Blake imagined this to be the scene as the fleet of the Mongol emperor of China, Kublai Khan, passed through the Indian archipelago in the thirteenth century. Kublai Khan's ships sailed across the seas of southeast Asia, plundering many settlements. The emperor's main fleet was decimated by hurricanes as it prepared to invade Japan.

↑ **Bringer of winter**
The ancient Greeks constructed this marble sculpture known as the Tower of the Winds in Athens. This side shows Boreas, who was said to bring the cold north wind. Many weather-related gods featured in Greek mythology; ancient Greece also saw the beginnings of scientific scholarship.

← **Lightning strikes**
This detail from a thirteenth-century illuminated manuscript, *De Natura Rerum* [Of the Natural World], depicts a cloudburst and three forks of lightning over the sea. For most people of the European Middle Ages, weather was a mysterious and often forbidding phenomenon, the vagaries of which determined their livelihood.

The

Meteorolo
cultural re
scientific r
the spirit c
instrumen
Galilei anc
realized th
changes in

THE AGE
The Renaiss
been adder
established
weather ac

GALILEO,
Galileo Gal
interest in
the stars th
thermome
inventor, b
of studying
convinced
phenomer
explained
brought h

Toward the modern era

The start of the twentieth century marked the beginning of the modern era of meteorology. During the early part of the century came a realization that there was a potential to forecast the weather on a variety of space and time scales. Lewis Fry Richardson (1881–1953) proposed using mathematical equations to predict weather. Although it was not feasible to do the necessary calculations at the time, the early years of the century saw the development of the first electronic computers, forerunners to those that today underpin the operations of national and international forecasting centers. Since then an extraordinary array of weather-observing techniques has evolved, from manual and automated observations to remote-sensing weather radar and satellites. Within a few decades, weather forecasting grew from being a flight of fancy into a reality.

EXPLAINING THE WEATHER WITH NUMBERS
Lewis Fry Richardson was the first to suggest that the weather could be predicted by numerical methods. He showed that weather and the atmosphere work together in a series of cascading scales, made famous through the celebrated ditty: "Big whorls have little whorls that feed on their velocity, and little whorls have smaller whorls and so on to viscosity."

Monitoring the sky In the 1930s, this United States National Guard aircraft was among the many airplanes adapted to make upper-air weather observations. The ability to construct weather charts for the upper atmosphere was a significant meteorological breakthrough. Upper-level pressure, temperature and wind speeds all have a strong impact on the weather patterns we experience at ground level. During the Second World War, for example, radiosonde networks expanded rapidly, with flights reaching well into the stratosphere. Hurricanes, tornadoes, thunderstorms and fronts could therefore be tracked much more accurately.

A computer ancestor The world's first general purpose electronic calculator—the Electronic Numerical Integrator and Computer (ENIAC)—was built in the late 1940s at the University of Pennsylvania, USA.

Radar and rain A radar at the UK Meteorological Office in the 1950s showed echoes from heavy rain or storms within a radius of about 150 miles (240 km). Developed during the Second World War, radar proved to be a major breakthrough in meteorology.

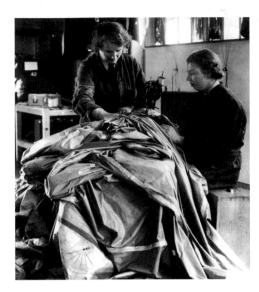

↑ **The wartime role of women** Women from the UK Women's Auxiliary Air Force repair a damaged weather balloon during the Second World War. Many women communicated weather information to support the war effort.

THE SATELLITE REVOLUTION

The era of weather satellites commenced with the launch of the first TIROS (Television Infrared Observation Satellite) polar-orbiting satellite in April 1960. These weather satellites were able to provide weather forecasters with a broad picture of cloud formations across wide expanses of the globe for the first time. Although the original satellites were simple instruments by today's standards, they were able to relay real-time cloud pictures to land-based forecasters. This marked the end of the era when fierce storms over the oceans could strike land before they were detected. Satellite meteorology advanced at an astonishing pace through the 1960s. By 1963 photographs could be obtained directly from satellites as they passed overhead. Three years later the first geostationary satellite, hovering over the equator, was launched.

→ **The first eye in the sky** Scientists prepare the world's first weather satellite, TIROS, for its launch in 1960. Since the success of TIROS there has been an ongoing international program of weather monitoring by orbiting satellites. This began in 1963, at the height of the Cold War, when 150 countries together set up the World Weather Watch (WWW) to exchange data on a regular basis and thus facilitate the preparation of global weather charts.

Meteorology today

Modern meteorology uses science and technology—most particularly meteorological satellites and high-speed computers—to monitor and forecast weather patterns. The preparation of accurate weather forecasts is aided by very high levels of cooperation between the national weather services—a level of mutual assistance probably unequaled in any other field. Operators of weather satellites, weather-station networks and computer models freely exchange information, all in close to real time. Researchers work closely together to help solve the lingering mysteries of weather and climate. And all of this collaboration is leading to predictions of increasing precision—forecasts for periods as short as an hour or as long as the next season or two. Whether you are drilling for oil over a remote ocean, planning a vacation or growing a grain crop, reliable weather information has an important role to play.

THE METEOROLOGICAL REVOLUTION

The practice of meteorology is reliant upon accurate data. The ability to rapidly communicate information around the world, process it, store it and display it before it becomes out of date—something which would leave practitioners of a century ago in awe—is the secret to the success of the modern meteorologist.

→ **Keeping track down south** Automatic weather stations, like this one in Antarctica, record such things as wind speed and direction, sunshine, temperature, humidity, atmospheric pressure and precipitation.

↓ **Prediction as an electronic image** A computer prediction of a weather system draws on a great many past records. The accuracy of computer predictions has improved dramatically as the power of computers has increased.

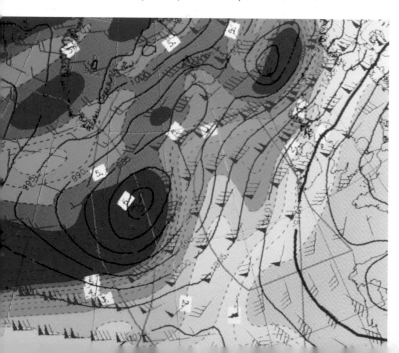

→ **Satellite's-eye view of a hurricane** Computer-enhanced imagery, such as this photograph of an intense hurricane over the Caribbean Sea, helps forecasters follow the movement of these potentially devastating storms. With computers, meteorologists can forecast their movement and issue warnings to communities at risk. The world's fastest supercomputers are already simulating constantly evolving weather patterns.

THE MODERN METEOROLOGICAL OFFICE

Meteorological forecasting centers have become high-technology offices with access to observational and forecast data only dreamt of a few decades ago. Meteorologists are able to follow developments in the weather on both small and large scales.

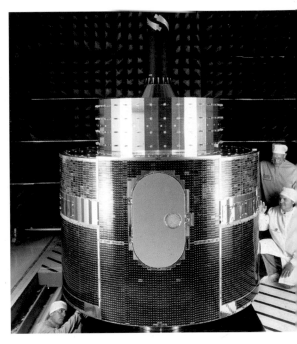

↑ **Spy in the sky** The Meteosat 6 satellite, here undergoing tests at the Aerospatiale facility in Toulouse, France, is typical of the satellites that provide almost instantaneous observations for forecasters. Since 1960, when the first, rather awkward, weather satellite TIROS was launched, the impact of satellites has magnified. The year 1966 saw the launch of the first weather satellite that orbited directly above the equator—an ideal orbit for monitoring severe weather.

← **Screen observations** A weather forecasting center at the United Kingdom Meteorological Office shows the heavy reliance on computer displays of current weather observations, satellite imagery and radar. Forecasts and warnings issued by these centers are distributed by high-speed communications, appearing within seconds on websites, mobile telephones, faxes and recorded messages.

Measuring weather

Networks of weather-recording stations have been set up in every nation, with some stations dating back 400 years. These thousands of weather stations provide data, usually every three hours, for national weather organizations. Ships and buoys carry instruments so that conditions over the world's oceans can be monitored, and automatic stations operate in isolated and inhospitable parts of the world such an Antarctica. Even the most basic stations measure temperature, humidity and rainfall. Many other elements can also be measured, with official weather stations usually recording the wind's speed, its direction and gusts, atmospheric pressure, sunshine, solar radiation, phenomena such as thunderstorms, hail, snow, fog and frost, cloud amount, type and height, and evaporation. Weather balloons are released at regular intervals and these provide details of the wind, temperature and humidity.

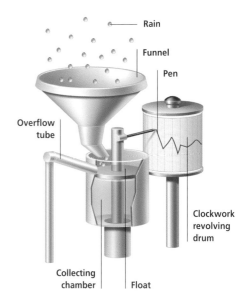

Rain

Funnel

Pen

Overflow tube

Clockwork revolving drum

Collecting chamber

Float

↓ **Auto-observation** Weather stations, such as this one at Valloire in France, are used around the world, to measure temperature, humidity, barometric pressure, and wind speed and direction.

↓ **Gauging the rain** Rainfall is possibly the most important weather element of all. Rain gauges collect all forms of precipitation, whether it is rain, sleet or snow.

MONITORING THE ATMOSPHERE

Weather observations are made and collected around the world on a continuous basis. Manually recorded data is retrieved at set times throughout the day and night, while information from automatic weather stations is transmitted to the various parent national weather stations as a constant flow of data. The observations are processed by meteorologists, who use the information to produce the weather maps that are reproduced in newspapers and on television.

In more recent times, mathematical simulation of the weather has become an increasingly important part of weather forecasting, and the observations taken from around the world are a vital part of this modeling process. With the network of automatic weather stations constantly expanding, mathematical simulations—and the accuracy of weather forecasting—are improving.

↑ **Putting rainfall on paper** This ingenious device, known as a pluviograph, measures not only the amount of rainfall but also the rainfall rate. The rain falls into a funnel and runs into a collecting chamber, causing a float to rise. The rise is measured on a revolving drum. When the water in the chamber reaches a certain height, the excess escapes through an overflow pipe. While professionals use an array of equipment to measure rainfall, natural signs can often provide early forecasts of impending rain: cattle gathering in a corner of a field; bees returning to their hives; frogs croaking; flowers opening and closing are among these signals.

Recording a typhoon A reconstruction of the barograph record produced by Typhoon Paka that passed near Guam on December 16, 1997 shows a trough of extremely low pressure that is indicative of a severe typhoon.

THE AEROSONDE

A fascinating new tool in the weather-watching arsenal is the aerosonde, a small, remote-controlled aircraft developed especially for long-range reconnaissance over oceanic and remote areas. It is being deployed to fill gaps in the global upper-air sounding network. With a 10-foot (3 m) wingspan, this aircraft can fly into hurricanes while they are still hundreds of miles out to sea, and accurately track their movements and intensity changes. The aerosonde can also track other weather systems across vast uninhabited areas and provide details of incipient storm development. With its arsenal of meteorological sensors, the robotic aerosonde could replace the surveillance now undertaken by specially strengthened military aircraft. It is one of many tools that will, in the future, revolutionize our ability to forecast the weather.

↑ **Traditional tool** Liquid-in-glass thermometers have been used for centuries to measure the temperature and humidity of the air. The vertical thermometers measure the temperature of dry and wet air; this allows the humidity to be calculated.

→ **The barograph** A barograph measures air pressure by utilizing a vacuum chamber—the silver cylinder near the center of the photograph. Changes in air pressure create changes in the height of the cylinder, which are transmitted to the horizontal arm at the front. This is fitted with a pen that draws a continuous line on the graph paper on the left.

↓ **Recording the daily ups and downs** A continuous graph of the air temperature can be recorded by this instrument, known as a thermograph, similar in principle to a barograph.

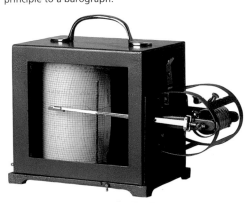

Weather at home

Observing the weather can be an intriguing hobby. Measuring and recording local weather provides not only an understanding of the immediate area, but an appreciation of how the elements interact on a broader scale. Some weather-watchers set up their own home weather stations—these can vary in complexity from just one or two basic instruments to sophisticated devices linked to a home computer that records and graphs observations. If a backyard is available, an instrument shelter can house and protect an array of instruments similar to the professional weather stations used by the local weather office.

THE BACKYARD WEATHER OBSERVER

Weather watching is an interesting and constructive hobby. A variety of high-quality meteorological instruments are available commercially, and these can be used in domestic situations to record weather developments in the immediate environment. Common instruments include rain gauges and thermometers, barographs (for recording atmospheric pressure), hygrographs (for humidity) and anemographs (for wind). The readings people obtain at home can be compared with those published in the press by the local meteorological office, or more frequently today, amateur weather-watchers can check their findings (and make links with other enthusiasts) via the internet. Weather phenomena are also perfect subjects for camera enthusiasts.

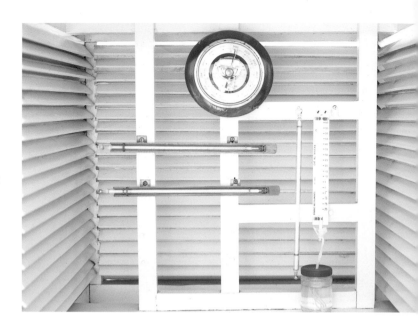

Home weather center A louvered timber shelter houses an aneroid barometer (*top*), wet and dry bulb thermometers (*left*) and horizontally mounted minimum and maximum thermometers (*right*). The slatted sides prevent sunlight and heat radiation from reaching the thermometers, while allowing a free flow of air.

← **All in one** This display combines an LCD calendar and clock, together with a thermometer and alarm. Instruments like these are inexpensive and make ideal gifts for amateur weather enthusiasts.

← **Barograph** A barograph is a fairly complex meteorological instrument which measures and records changes in atmospheric pressure. It has to be precision-made to function accurately. A pen, attached to a pointer, plots the pressure changes on cylindrical graph paper—unlike most barometers, in which the pointer is on a dial.

Gauging the rain A rain gauge—used for measuring how much rain has fallen over a given period—is one of the simplest instruments available. It should be placed well clear of trees and buildings.

Three in two This instrument has two dials but displays three readings: at top is an aneroid barometer for measuring pressure; below temperature and humidity readings are shown.

Triple reading Some instruments have multi-function display panels. This device measures and displays humidity, temperature and barometric pressure readings.

LOCAL AND GLOBAL

Professional meteorology is becoming increasingly sophisticated. But there is still a place for the amateur weather watcher who wants to understand local weather conditions and how they fit into a wider perspective. The best way to understand local weather is to gaze at the sky and follow its ever-moving display.

← **Right weather, right plants** An understanding of local weather conditions enables professionals and amateurs alike to plan their plantings to ensure maximum success.

Monitoring the Sun

The growing popularity of solar power since the 1970s has produced a demand for meteorologists to produce so-called sunshine data—information that includes the "strength" of the Sun's rays, together with the average number of hours of sunlight at various locations and times. This data is vital for the efficient installation of solar-power technology, particularly the siting of energy-collection devices such as solar panels. Sunshine data is now routinely collected from many sites, often by national weather services. Many private organizations and research institutions such as universities also collect data on sunshine. Basic Sun-monitoring equipment is so simple to operate that many amateur weather watchers acquire it as a valuable and interesting addition to a home weather station.

MEASURING THE SUN'S RAYS

There are several different ways of measuring the Sun's rays. Complex electronic devices called pyranometers accurately monitor several aspects of solar radiation, while simpler instruments going under the generic name of sunshine recorders provide information about the number of hours of sunshine.

↓ **Solar exposure** Solar-recording devices such as these are positioned well away from any overshadowing objects. Solar radiation is related to the "power" of the Sun's rays and differs from region to region, as well as between seasons. Interest in solar power has spurred the development of sophisticated monitoring devices.

↗ **The Sun makes its mark** A weather observer notes the amount of sunshine measured by a Campbell Stokes sunshine recorder. By using a magnifying glass to produce a burn mark on a card, this instrument records the day's hours of sunshine. This traditional way of monitoring the Sun is gradually giving way to electronic recording systems.

THE SUNSHINE RECORDER

The main feature of the Campbell Stokes sunshine recorder is a glass sphere that acts as a magnifying glass. The glass concentrates the Sun's rays on a card behind it to produce a scorch mark. As the day progresses, the burn moves along the card, eventually providing a continuous scorch mark which allows meteorologists to calculate the number of hours of bright sunlight. If cloud obscures the Sun, there is a break in the scorch mark. While the traditional Campbell Stokes sunshine recorder is suitable for home weather stations, increasingly sophisticated electronic versions provide more accurate data.

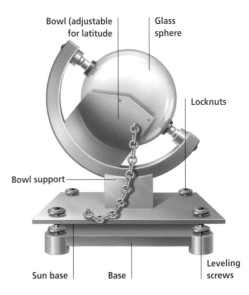

Bowl (adjustable for latitude)

Glass sphere

Locknuts

Bowl support

Sun base

Base

Leveling screws

CREPUSCULAR RAYS: IT'S ALL AN ILLUSION

Crepuscular rays are a common phenomenon involving the Sun's rays and can produce quite beautiful and sometimes awe-inspiring effects as they streak across the sky. The phenomenon occurs when the Sun's rays are made visible by haze or dust in the atmosphere; to observe them properly the Sun has to be at least partially obscured. The Sun may be behind a cloud or it may be slightly below the horizon. Although the crepuscular rays appear to radiate outward from the Sun in a spectacular fanlike fashion, they are, in fact, parallel, with the divergent effect being an optical illusion. The same illusion is provided by a pair of railway tracks that run to the horizon where they appear to join together. When the Sun is low, toward dusk, the light rays often appear reddish and the dark shadows may have a greenish tinge. With a high Sun, rays may be seen penetrating breaks in stratocumulus clouds. In rare cases, crepuscular rays extend over the entire sky.

→ **Sunrays made visible** A view of crepuscular rays taken in California, USA, shows fine streaks of light and shadow beaming dramatically upward from the Sun, which is obscured just below a mountain range. Crepuscular rays are also known colloquially as Jacob's Ladder and Buddha's Fingers.

↘ **Sundeck** Meteorologists service a group of sophisticated solar-recording instruments at a solar-radiation research facility in the United States.

↓ **Sun car** This car has solar panels to convert sunlight into extra power to supplement its battery.

Weather organizations

Most countries have a national weather service to provide the large range of weather information required by their populations. The services issue forecasts and warnings, and provide local weather observation programs, climate services, environmental education and a variety of consultative activities.

No two weather services are precisely the same, although there is much common ground. Because of the global nature of weather, international cooperation is essential, and observations, forecasts and warnings are mostly exchanged free of charge between nations. This cooperation is encouraged and supported by international endeavors such as the World Meteorological Organization (WMO) and World Weather Watch (WWW).

ON-SITE OBSERVATIONS
Mount Washington Observatory in New Hampshire, USA, is a scientific and educational institution that maintains a permanently staffed observatory at the mountain's summit. The observatory conducts observational, research and educational programs that have provided valuable information about high-altitude environments

→ **Frozen station** The bitter cold experienced at the summit of Mount Washington is obvious from this photograph. Data collected from such remote localities provides a valuable insight into the extremes of weather.

← **Weather station** A meteorologist at the Weather Centre in Hong Kong accesses information not only from local sources, but also from a large number of international data banks. This global approach to the weather is essential to further our understanding of Earth's atmosphere. New technologies will lead to improved predictions, but one advance may well be in our understanding of forecasting limits.

↓ **Satellite power** A satellite view of the area around St. Louis, USA, at the height of the flooding of the Mississippi and Missouri rivers in 1993. The city is the purple area near bottom center. Blue and black areas show the extent of the flooding. The Mississippi River runs from top left to bottom right; the Missouri River flows from center left. Cooperation between satellite-operating countries improves the dissemination of information such as this.

THE WEATHER KNOWS NO BORDERS
The World Meteorological Organization, or WMO, was established in 1951 as a specialized agency of the United Nations. It has over 180 member countries, and acts as the world's "umbrella" meteorological organization, providing what it describes as "the authoritative scientific voice on the state and behavior of the Earth's atmosphere and climate." The WMO coordinates global scientific activity to facilitate prompt and accurate weather information and work toward a uniform weather-reporting system and better dissemination of meteorological information. Through its activities the organization aims to contribute to the safety of life and property, the socioeconomic development of member nations and the protection of the environment as a whole.

FLYING INTO CLEAR SKIES

The aviation industry is one of the world's most demanding users of weather information. History is full of examples of aircraft that have fallen foul of wild weather but nowadays, thanks to improved forecasting techniques, such incidents are becoming relatively rare. Pilots of modern airliners select their flight paths according to the forecast winds and tailor their fuel loading according to the expected weather delays en route.

→ **Forecasting and the military** Military operations on land, at sea and in the air are very sensitive to the weather. Military strike aircraft such as this one have a severely limited fuel supply, making accurate landing forecasts essential. Indeed, throughout history battles have been won and lost at the whim of the weather.

Weather models

At the beginning of the twentieth century scientists believed it was possible to predict how the weather would behave using mathematical equations. Unfortunately these equations were so complex that the experts were unable to prove their theories. The invention of the computer solved this problem, and computerized weather models of ever-increasing complexity and detail are now run by national weather services several times each day. The predictions from these computer models have become one of meteorologists' most powerful forecasting tools. Some provide a lot of detail over limited areas for a one- or two-day prediction; others provide less detail but their predictions may be for one or two weeks. More advanced computer simulations, which can process ever-increasing amounts of data, are in development.

TAKING EVERYTHING INTO ACCOUNT

Weather models use a large number of mathematical equations to calculate how the atmosphere and oceans are likely to behave. Some equations describe how the air flows in three dimensions; others predict the formation, movement and decay of clouds of all types, along with the likelihood of them producing rain, snow and hail. And still more equations are used to simulate all the other variables that influence the weather's formation. Mathematical modeling improves year by year as the number of worldwide weather observations increases and evermore powerful computers are developed.

↑ **Electronic assistance** The huge number of calculations used in the compilation of a computer weather model requires the largest and fastest machines available. Supercomputers such as this one are used in major weather-modeling centers. In 1922 British mathematician Lewis Fry Richardson first applied mathematical techniques to produce a crude 24-hour forecast, a forerunner of numerical prediction techniques. Today, ensemble modeling techniques, which use multiple computer runs to provide a range of possible weather outcomes, identify the state of the atmosphere. Such techniques will continue to be refined.

↓ **Color-coded atmosphere** The combination of bright colors, solid lines and arrows allows this map to display several weather phenomena at once; this enables forecasters to see how the different factors in the pattern interact.

↓ **The wind, seen from above** Forecasters need to see how upper-level wind patterns vary in relation to surface patterns. This three-dimensional display, with arrows representing wind speed, shows how wind patterns change with height.

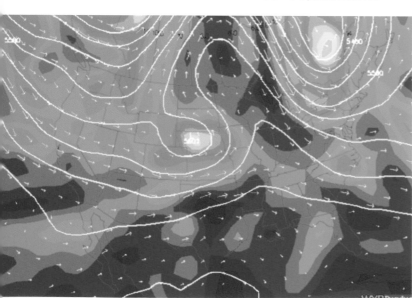

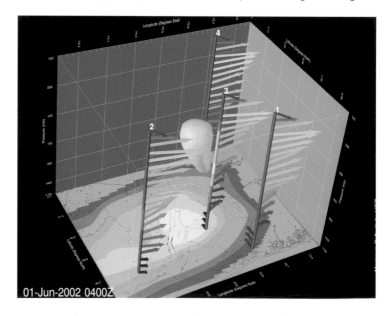

01-Jun-2002 0400Z

WHAT IS MESOSCALE MODELING?

A new term is creeping into the language of weather forecasters. Mesoscale modeling refers to the use of computers to predict features that highlight variations in the weather within states and even between neighboring countries in more detail than was available from earlier models. These models also provide predictions at quite brief intervals, enabling meteorologists to improve their forecasts of the time of onset and clearance of weather phenomena such as rain, snow and strong winds.

→ **"Seeing" the weather** Modern technology allows meteorologists to generate a three-dimensional, multi-colored image of a thunderstorm.

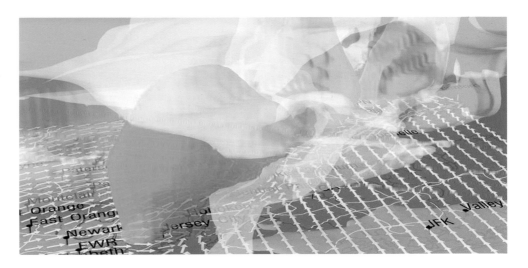

MODELING A THUNDERSTORM

Sophisticated computer visualizations are widely used to display model predictions. The mesoscale display above, originally in a three-dimensional format, shows severe thunderstorms on a frontal weather system over the southeast of New York State, USA, on May 31, 2002. The white areas are cloud and the pale blue areas the strongest parts of the thunderstorm. Such models are used to investigate the causes of unusual and severe weather patterns; their findings are employed in models used to create more accurate and timely day-to-day weather predictions.

Wind "ribbons" This model uses ribbons to show how winds flow into a low-pressure system then into the upper atmosphere.

Seasonal forecasts

From the earliest days communities have made decisions about many of their activities based upon what the climate "normally" brings for a particular time of the year. Unfortunately the weather does not often conform neatly to an average level. An improved understanding of what can be expected over a coming season is of immense value to farmers. Several techniques are used by meteorological centers in the preparation of seasonal forecasts. Many are based upon statistics that link past wet or dry, or hot or cold, events with various predictors, such as, for example, persistence—what has been happening will continue to happen. More skilful techniques correlate future weather conditions with sea-surface temperature anomalies in oceans or the extent of pack-ice in polar regions, both of which produce effects that may last for months or even years.

LOOKING SEVERAL MONTHS AHEAD

Computer models are used extensively to study how the weather and climate change on both global and regional scales for several months ahead. For these long-range predictions, the computer models must simulate how both the atmosphere and oceans behave. Scientists are also realizing that it is important to understand how the world's large ice sheets, as well as atmospheric pollution and changes in global vegetation, contribute to longer term climate changes.

→ **The regional view** Significant temperature variations can be seen in this summer 2002 regional prediction for the USA. With regional models being employed in making a seasonal prediction, the valuable result is greater regional detail.

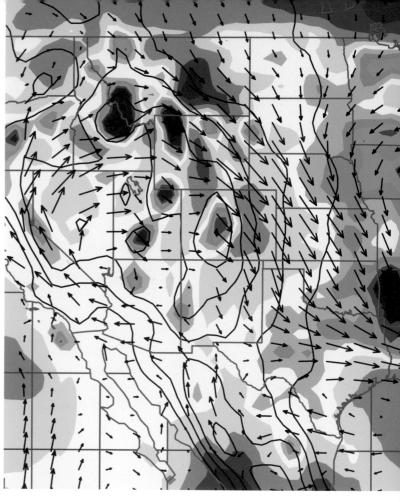

A PREDICTION FOR ALL SEASONS

Weather forecasts for the next day or two are usually fairly specific, particularly the forecasts of maximum and minimum temperature. However, a forecast for a whole season makes extensive use of probabilities. This is because such a long-range forecast is far less certain than a short-range forecast. Multiple weather systems will affect any given region in the course of a season. The seasonal forecasts attempt to combine the effects of all of these weather systems for the following months. Predictions then provide the most likely probability of whether temperatures, rainfall, winds and humidity will be greater than, equal to or less than normal for different regions for the entire period, rather than trying to identify the effects of individual weather systems.

← **The wind's impact** A summer 2002 prediction foresaw northerly winds in the central USA and heavier than normal rainfall over the eastern states. Predicted long-term temperature and precipitation variations may often be explained by expected wind anomalies.

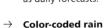

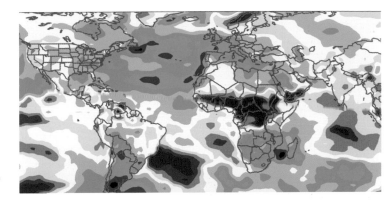

→ **Seeing into the near future** In this seasonal temperature forecast for the northern hemisphere summer in 2002, red patches indicate warmer than normal temperatures with blue shades depicting regions likely to be cooler. Sophisticated seasonal predictions may become as common as daily forecasts.

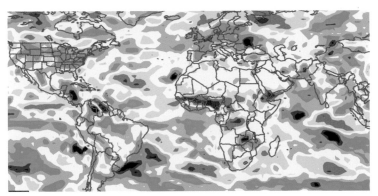

→ **Color-coded rain** Knowledge of future rainfall is vital in many industries. In this three-month rainfall forecast for the northern summer of 2002, blue represents wetter than normal conditions while red indicates drier weather. Forecasts for summer and winter are more accurate than those for spring and autumn.

THE OCEAN'S IMPACT ON THE ATMOSPHERE

In recent decades scientists have come to realize that changes in the distribution of "pools," or large areas of warm and cool water in oceans around the world, have a major influence on what the weather will be in the ensuing months. There is a tremendous amount of energy contained in the top layers of the oceans. When the water is warm, there is more energy available to transfer to the atmosphere. Warm waters also send more moisture into the air—moisture that can link up with suitable weather systems to produce increased rainfall. Cooler water has the opposite effect.

Weather patterns can be affected around the globe when these sea-surface temperature anomalies cover tens of thousands of square miles of the ocean surface. The El Niño and La Niña weather phenomena are strongly tied to these anomalies.

→ **Aid to farmers** This forecast includes predictions of soil moisture anomalies; drier than normal soils are in red, wetter than normal soils are shown in blue. Farmers may use such forecasts in planning future plantings

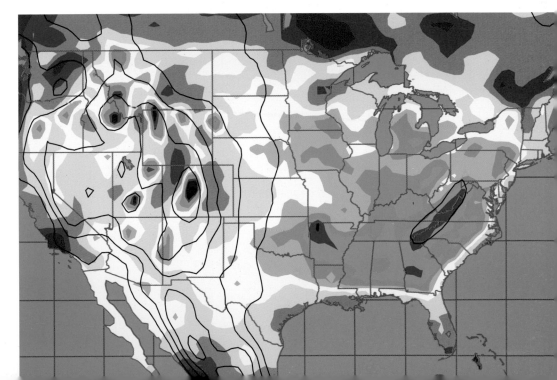

Mapping the weather

Weather maps, known to meteorologists as synoptic or prognostic charts, are the main tools of meteorologists. Synoptic charts provide a snapshot of an area's weather at any given time; prognostic charts predict likely future weather. Weather observations form the starting point of all charts. These are taken at the same, set times around the world so that there are no discontinuities in weather maps from one time zone to the next. The information from land and sea is transmitted to forecasting offices, where it is plotted as symbols on large charts to consolidate all the data.

THE PUBLIC WEATHER MAP

Weather maps are now more accessible to the public than ever before. In their daily news programs, television stations display simplified charts, some with colorful and animated visual effects. Most newspapers publish weather maps, sometimes for several days ahead. The internet too has a large range of weather maps, most of them computer-generated.

↑ **Seeing the pattern in it all** A meteorologist uses satellite imagery to help position frontal systems on his chart. Drawing charts is how meteorologists keep track of weather systems that affect the weather across their regions. Weather maps are the result of condensing a huge amount of information into an internationally recognized, format.

← **Yachtsman's nightmare** Blue-water ocean races can put yachtsmen in the path of high seas. Access to current weather forecasts and warnings through satellite and radio communications helps sailors steer a safe course. Radio and communication networks provide regularly updated bulletins for sailors.

↑ **Tracking a front** Weather maps show cold fronts, such as this front passing to the south of Africa, as lines with triangular barbs on them. This enables forecasters to track their movement accurately across the oceans.

→ **International weather symbols** A set of internationally accepted symbols enables meteorologists to interpret data on weather maps.

↓ **Making "weather" visible** This map, like most weather maps, uses isobars—lines that join places of equal air pressure—to show a deep low-pressure system over the English Channel. High-pressure systems lie to the west, south and east.

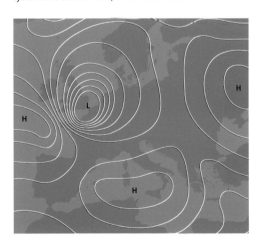

INTERNATIONAL WEATHER SYMBOLS

Current weather

light drizzle	steady, moderate rain	hail
steady, light drizzle	intermittent, heavy rain	freezing rain
intermittent, moderate drizzle	steady, heavy rain	smoke
steady, moderate drizzle	light snow	tornado
intermittent, heavy drizzle	steady, light snow	dust storms
steady, heavy drizzle	intermittent, moderate snow	fog
light rain	steady, moderate snow	thunderstorm
steady, light rain	intermittent, heavy snow	lightning
intermittent, moderate rain	steady, heavy snow	hurricane

Low clouds

stratus	cumulus	cumulonimbus calvus
stratocumulus	cumulus congestus	cumulonimbus with anvil

Middle clouds

alstostratus	altocumulus	altocumulus castellanus

High clouds

cirrus	cirrostratus	cirrocumulus

Sky coverge

no clouds	four-tenths covered	seven- to eight-tenths covered
one-tenth covered	half covered	nine-tenths covered
two- to three-tenths covered	six-tenths covered	completely overcast

Wind speed mph (km/h)

calm	9–14 (14–23)	55–60 (89–97)
1–2 (1–3)	15–20 (24–33)	119–123 (192–198)
3–8 (4–13)	21–25 (34–40)	

Tracking the weather with radar

For short-term forecasting, nothing can quite match the detail and versatility of weather radar. Modern weather-radar networks are able to produce detailed, three-dimensional images of weather systems—from thunderstorms to fronts to hurricanes—every five to ten minutes. For this reason, weather radar has become an indispensable item in the meteorologist's toolbox. It is the only instrument that can alert forecasters to the likely formation of severe thunderstorms and tornadoes. Radar also permits national weather centers to follow accurately the eye of hurricanes as they bear down upon coastal communities. Live radar imagery is now available on the internet and television weather presentations, enabling the general public to be continuously updated on where severe weather is occurring. Regions with good weather-radar coverage now benefit from warnings with a speed and precision not thought possible in the early days of weather forecasting.

SCANNING THE SKIES

Weather radar emits pulses of radio waves that bounce back from various bodies, or "scatterers" in the air. Although smoke and dust can scatter radar signals back, the strongest echoes come from raindrops, snow and hail. A Doppler radar measures the change in frequency of the returned echoes, enabling wind speed to be calculated. These devices have a very narrow beam which scans the skies at different angles so that a three-dimensional picture of the weather is produced. Radar can peer deep into approaching thunderstorms.

← **Flying into the storm**
Aircraft such as this radar-equipped Electra L188L go as close as possible to the heart of developing storms to provide scientists with three-dimensional detail of the wind and rainfall patterns within. The United States has by far the most comprehensive network of weather-watching radar, with almost total coverage of the mainland. In a country where 10,000 violent storms, 5000 floods and up to 1000 tornadoes strike each year, this extensive radar coverage enables more accurate identification of potentially damaging storms, a reduction in false alarms and an increase in warning times. In coming decades, dual polarization techniques, which identify hail suspended inside clouds, will move from research into operational use and further increase radar's sensitivity and accuracy.

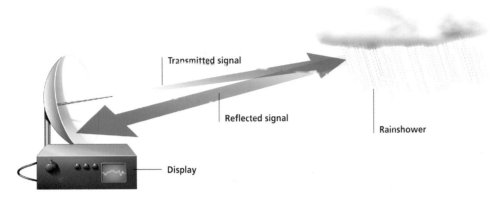

Transmitted signal

Reflected signal

Rainshower

Display

HOW WEATHER RADAR WORKS

Radar, an invention that appeared during the Second World War in order to detect incoming enemy aircraft during the Battle of Britain, stands for "radio detection and ranging." It was realized that it could also detect rain and so it soon became an indispensable meteorological tool.

When a radio beam is emitted from an antenna and meets a "target" such as a shower of rain, some of the beam is reflected back. This reflected beam is displayed on a screen that shows the range and distance of the "target." Meteorologists know where the nearest rain to an antenna is located.

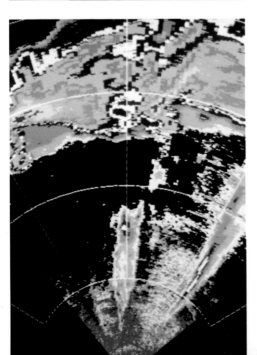

↖ **Turning radar pictures into a forecast**
A scientist studies airborne-radar imagery and other data collected as part of the FASTEX (Fronts and Atlantic Storm Tracks Experiment) program.

← **Snapshot of a tornado in the making**
A real-time radar image of a severe thunderstorm showing "hook" echoes is one indication that a tornado is forming.

↑ **Looking into the hurricane's heart** A weather-radar image superimposed on a satellite image produced this picture of Hurricane Hugo as it crossed the coastline of South Carolina, USA, in 1989; the eye of the hurricane can be clearly seen. Combining different views of the same phenomenon helps meteorologists distinguish the important features of hurricanes or storms and provide timely warnings.

Weather from above I

Following the launch of the first meteorological satellite in 1960, there has been a steady increase in the numbers of these "eyes in the sky." Detailed photographs of the cloud patterns far below have been invaluable to meteorologists, and the images have generated considerable popular interest. The wealth of visual data from unmanned space vehicles, as well as manned space flights such as the space shuttle and space station missions, has significantly increased our knowledge of the mechanics of the atmosphere and the powerful forces that create storms. Although some of the older images are now more than 40 years old, their perspective of our planet makes them as relevant now as when they were first taken.

PORTRAITS OF EARTH

Early space photography was in black and white, but colored images arrived with the launch of manned flights in the mid-1960s. As the technology improved, so did the quality of the imagery, and today high-resolution photographs of astonishing clarity are available. This enables meteorologists to see fine-scale cloud patterns that were previously invisible.

↓ **Where the trade winds blow** The extensive cloud formations that drift along the mountaintops of the Hawaiian island of Oahu are the result of the northeast trade winds encountering the ranges. The winds are blowing from the top right of the photograph toward the bottom left and clouds are developing along the windward slopes of the Koolau Range. Lush vegetation can be seen on the high ground, and dry rain shadow on the leeward side.

↑ **End of a thunderstorm** The circular cloud pattern occurred after a thunderstorm collapsed. Squally, outflowing winds are the result, spreading out radially from the previous location of the thunderstorm's center.

↓ **Hurricane offshore** A September 2001 photograph shows Hurricane Erin spinning north of Bermuda. The Great Lakes are visible toward the top left.

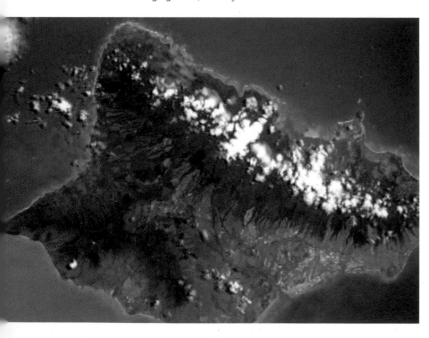

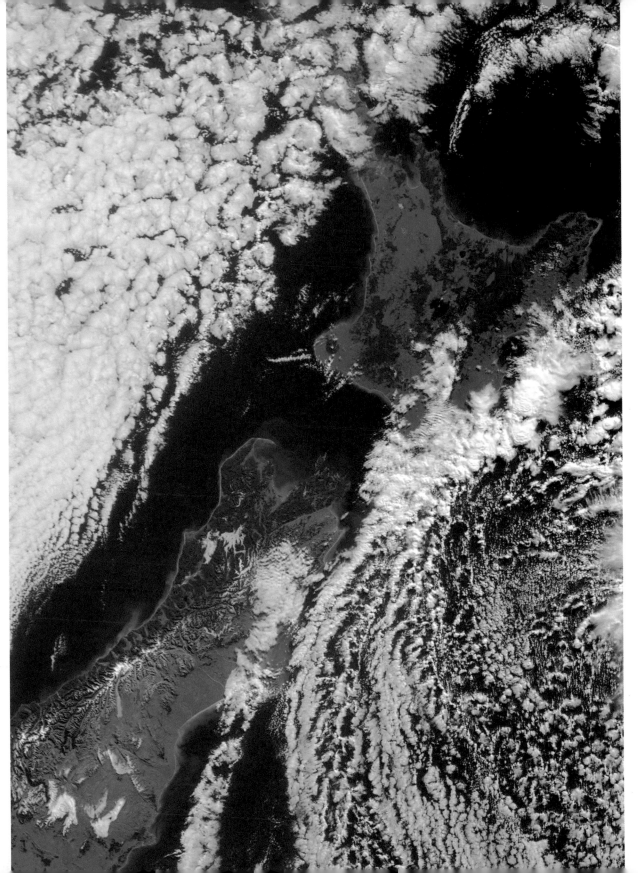

THE ORBITING WEATHER SPIES

Modern high-resolution space photography allows us to see intricate cloud patterns that were unknown until a few decades ago. Satellite images have made it possible to see the position of single cloud cells, and as a result meteorologists have a much-improved insight into the formation and development of small-scale weather systems. These include sea breezes, thunderstorms and von Karman vortices, which are the intricate cloud patterns formed when wind flows past an obstacle, such as an island or a mountain. In addition to photographing clouds, many satellites are able to provide information on the state of the sea, upper-level winds and surface temperatures. The photographs taken from space have another value, too: they provide a unique perspective of our planet and its place in the cosmos.

New Zealand revealed New Zealand's two islands are framed by narrow cloudbands—the result of winds from the southwest forcing a large cloudmass to part at the South Island's southern tip.

Weather from above II

The first weather satellites typically orbited at an altitude of 500 miles (800 km) above Earth. Later, from 1966, came high-altitude satellites that sit some 22,230 miles (35,800 km) above the equator and orbit Earth at the same rate as the planet itself. These so-called geostationary satellites provide a series of images that create a moving picture of clouds as they develop and dissipate. This is of vital importance in measuring and forecasting cold fronts, hurricanes and low-pressure cells.

→ **Yangtze in flood** When a flood surged down the Yangtze River, China, in August 2002, the embankments made by local Hunan farmers held. The top image shows the region before the flood; the bottom image shows the extent of the inundation.

↓ **Hurricane over Mexico** This MODIS image shows Hurricane Douglas off Baja California, Mexico, in July 2002. Photographs of this kind are key resources in forecasting and measuring weather phenomena.

THE NEXT GENERATION
A new generation of satellites was launched in the 1990s. The United States' Earth Observing System satellites carry the MODIS package, a key instrument in monitoring climate change by observing sea ice, glaciers and oceans. Another US space vehicle is SeaStar, which carries the Sea Wifs package. This monitors the role of the oceans in determining Earth's climate.

AN INTERNATIONAL INITIATIVE

The geosynchronous orbit—directly above the equator—is ideal for measuring severe weather conditions as it permits the continuous tracking of ever-changing storms. Weather satellites in geosynchronous orbit are increasing in number and complexity. Today there are seven geosynchronous satellites, operated by the United States (two), Russia, China, Japan, India and the European Union. Technical advances in these satellites will provide clearer, more realistic images that will allow meteorologists to provide more accurate short-term and medium-term forecasts. There is widespread cooperation between the world's satellite-operating nations, and programs are coordinated to maximize satellite coverage.

Natural inspiration I

Throughout history, artists have been inspired by the ever-changing natural displays provided by the weather. Many decades before the development of satellite photography in the 1960s, the nineteenth-century artist Vincent van Gogh was depicting turbulent, swirling cloud formations in his extraordinary paintings. Although his expressive clouds may seem to have little relation to an Earth-bound view of the sky, they do bear an arresting resemblance to the images of Earth's atmosphere that have been captured from satellites. Indeed, the remarkable patterns, shapes and colors of satellite photographs can be as captivating as masterpieces of modern art. Science and art appear to be inextricably linked.

↑ **Swirling clouds**
Van Gogh painted *Mountainous Landscape Behind Saint-Paul Hospital* in early June, 1889, while staying at an asylum in Saint-Rémy-de-Provence, France.

← **Sky in tumult** *Starry Night*, possibly van Gogh's most famous painting, was also painted in June 1889. The tumultuous sky has often been interpreted as a reflection of the artist's troubled mind. There has been conjecture that the stars are those of the constellation Aries and the planet Venus (the brightest starlike object at top right). Venus was in a crescent phase when the painting was made.

→ **Aerial pattern**
Landsat 7 took this image of clouds over the western Aleutian Islands of Alaska. The color variations were created by differences in temperature and in the size of the clouds' water droplets.

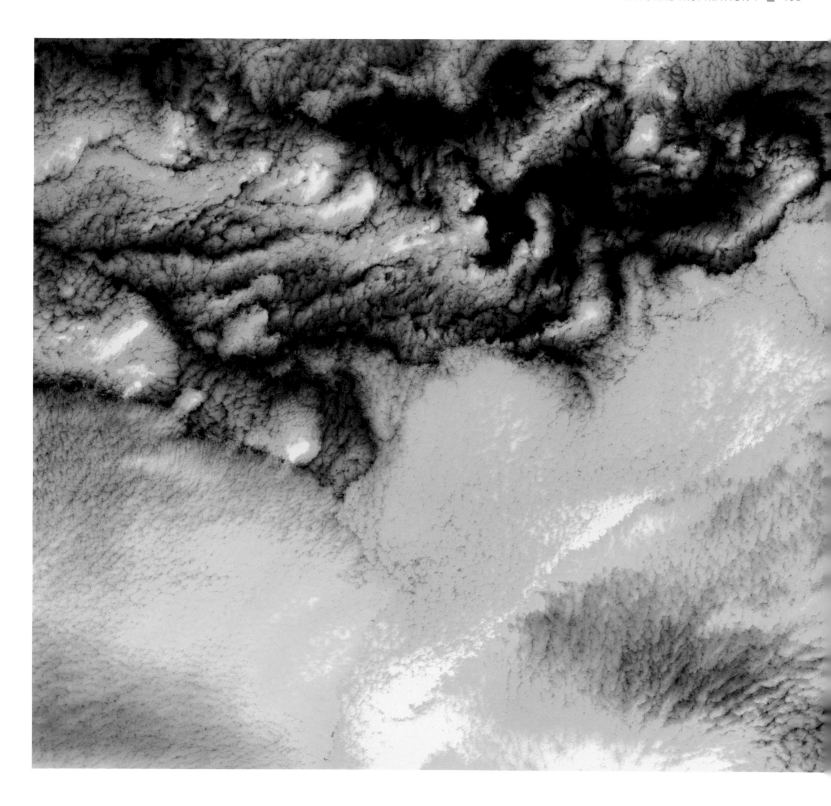

Natural inspiration II

In recent times, so-called mathematical art has gained popularity. This involves expressing the outcome of a mathematical process as a color rather than a number. In some cases this produces images startlingly similar to spiraling cloud formations, particularly those viewed from space. Spiral patterns are also found in radar images, where they are caused by the reflection of the radar beam off water droplets.

→ **Radar work of art** A radar image shows spiral-shaped bands of precipitation associated with a low-pressure cell approaching the Californian coast. The spiral is a phenomenon of atmospheric processes as well as artistic expression.

↓ **Waves in the Weddell Sea** Taken from the space shuttle Endeavour, this photograph shows two large eddies visible at the northernmost edge of the pack ice. These patterns are strikingly similar to cloud patterns commonly visible in the atmosphere, and the whorls and eddies can also be replicated by computer-generated mathematical art.

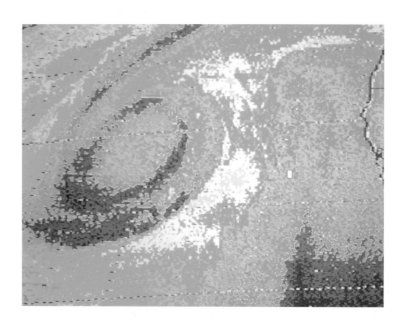

FRACTAL WHORLS AND EDDIES

What is referred to as fractal geometry is an unexpected byproduct of the computer age that has made a major contribution to the field of mathematical art. It employs a huge number of mathematical calculations, the end result of which is shown as a color located precisely on a graph. Before computers it was impossible to produce images of this nature because of the millions of calculations that were necessary. One of the obvious features of fractal geometry is its similarity to naturally occurring phenomena, such as clouds, lightning and various types of fluid interaction. In fractal geometry, as in nature, there are few straight lines and the environment consists instead of countless spirals, whorls and eddies that can produce a canvas of intriguing beauty and complexity.

Abstract beauty This dramatic image is an example of fractal geometry presented as mathematical art. It is called *Diamond Mine* and is strikingly similar to some cloud pattens as viewed from Earth-orbiting satellites.

Global climate

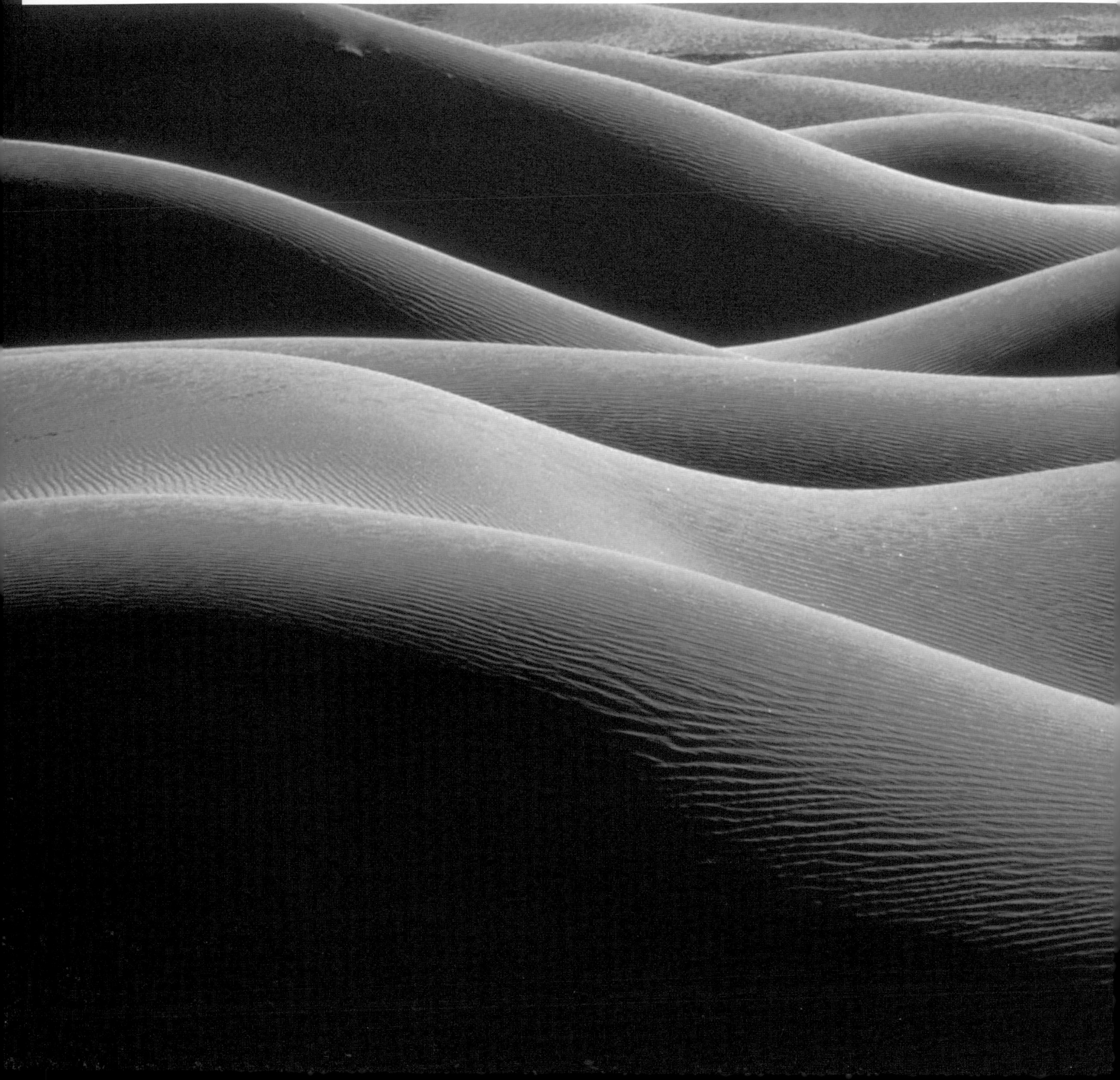

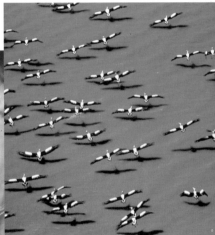

Previous page Desert sand dunes are among Earth's most inhospitable habitats.

Global climate

From Antarctic emperor penguins to the zebras of the African savanna, from lush tropical forests to desert cacti, the great diversity of life on Earth is largely a response to the challenges presented by our planet's different climates. Throughout the globe, humans too have made the most of this climatic variety.

Climate zones

By 500 BC, the Greeks had classified the world's climates into torrid, temperate and frigid zones according to the angle of the Sun. During the early twentieth century, a comprehensive climate classification scheme was developed to coincide with global vegetation patterns. Based on a locale's monthly temperature and precipitation averages, this scheme is still widely used today.

The climate of a location depends upon its latitude, altitude, and distance from large water bodies and mountain barriers, as well as its relationship to global atmospheric circulation patterns. While the generally accepted climate classification used in this book provides a broad overview of global climate, there are significant anomalies within regions that result from local weather patterns. In turn, the incredible variety of life on Earth is largely due to climate, as plants and animals have evolved to thrive in the many different conditions.

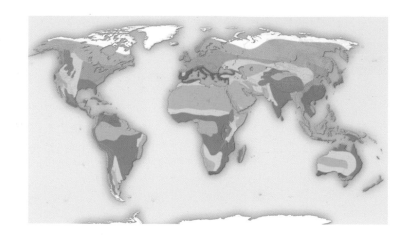

WORLD CLIMATES
The map of global climate zones shows a distribution related primarily to latitude, with regions broadly defined by the tropics of Cancer and Capricorn and the Arctic and Antarctic circles but also affected by altitude and proximity to oceans.

↓ **Arid** Arid climates have little moisture in the air, leading to scant precipitation and pronounced daily and seasonal temperature variations. Because there is little vegetation, sand grains are easily collected by the wind and piled into dunes, which are in a constant state of motion.

CLIMATE ZONES

- Tropical
- Subtropical
- Arid
- Semiarid
- Mediterranean
- Temperate
- Northern temperate
- Polar
- Mountain
- Coastal

Mountain A mountain climate, such as is found at higher elevations in the French Alps, has temperatures that are lower than those in the valleys below. Winter snowfall can be significant and winds can be fierce. Plants and animals must also cope with the stress of reduced levels of oxygen.

→ **Tropical** The high temperatures, rainfall and humidity of the tropical climate are responsible for the lush vegetation of South America's Amazon Basin. Rain forests teem with plant and animal life: more than 1.5 million species have been recorded to date.

↓ **Temperate** A temperate climate's seasonal cycle is reflected in the fall colors of deciduous trees, such as maple, birch, oak and hickory, which drop their leaves and become dormant in winter.

↓ **Coastal** The weather in a coastal zone, such as the rocky shores of Thailand, depends on the local sea-surface temperatures, but air temperature variations are small compared to neighboring inland locations. Although the temperature is relatively stable, winds, waves and salt spray can be harsh on vegetation. Plants tend to grow close to the ground, and many have waxy leaves that prevent water loss and large root systems to anchor them in unstable sand.

Polar Antarctica has Earth's coldest, driest and windiest weather. Extended intervals with no sunlight make the winters extremely cold and long, while sunny summers remain cool and relatively short. Penguins and other animals can survive only in the milder polar zones along the coast or on the Antarctic Peninsula. In the Arctic, only the strongest animals, such as polar bears, musk oxen and wolves, can survive the winter in the open; others hibernate or migrate to warmer climes.

Tropical

Tropical climates are found in the equatorial regions that lie between the tropics of Capricorn and Cancer, and feature high temperatures and substantial precipitation throughout much of the year. With little seasonal variation in the intensity of overhead sunlight, temperatures remain high, with the lowest monthly temperature no lower than 65°F (18°C). While there can be rainy and dry seasons, rainfall is typically high, with at least 4 inches (100 mm) of rain every month. The seasonally constant light, warmth and rainfall of tropical zones provide ideal conditions for life, giving rise to lush rain forests and producing the greatest species diversity on Earth. Tropical rain forests occupy only 7 percent of Earth's landmass but contain 50 percent of the world's plant and animal species.

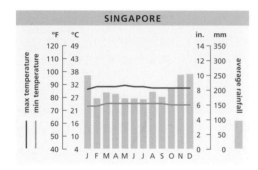

SINGAPORE

HOT AND WET Singapore shows little temperature change throughout the year, with the minima and maxima varying by only one or two degrees. Rainfall is high all year, with a peak around November to December. Uncomfortably high humidity is common.

Reaching for the light Because the climate allows continuous, rapid plant growth, space is at a premium in tropical rain forests. Vines such as lianas use tree trunks as support in their race for the light.

The humid zone The humidity of the rain forests of Central and South America allows the poison-arrow frog to exploit both land and trees as habitat. The bright colors warn predators that the frog is toxic.

Life in the forests Tropical primates such as the white-faced capuchin monkey are adapted to life high in the forest; they have superb stereoscopic vision and rotating hands for grasping branches.

Rain forest layers Luxuriant tropical rain forests support a broad diversity of fauna in four distinct habitats—the emergent layer with the tallest trees, the canopy forming a dense cover of tree crowns, the dim understory and the moist forest floor.

WORLD TROPICAL ZONES

The tropical climate zone straddles the equator, with extensions reaching the tropics of Cancer and Capricorn. It encompasses parts of Central and South America, including the Amazon Basin; Africa's Congo Basin; parts of Southeast Asia, including Indonesia and the Philippines; and parts of Oceania. Day length changes little throughout the year.

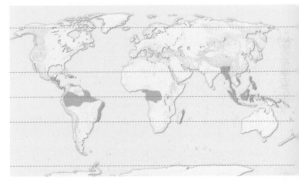

Feathered display Brilliantly colored birds such as the scarlet macaw live in tropical forests. Like other parrots, the macaw has a strong bill that can crush nuts and seeds, and dextrous feet for climbing trees.

Rainwater reservoir The bromeliad is an epiphyte, a nonparasitic plant that grows on trees and obtains nutrients and moisture from the air. The long leaves funnel rainwater into a reservoir at its base.

Amazon hunter The semiaquatic caiman lizard lives in the Amazon Basin. It searches the waterways for mollusks such as snails and mussels, crushing their shells with its powerful jaws.

Subtropical

Found across tropical and subtropical latitudes, the subtropical climate features a distinct wet season/dry season cycle with relatively high temperatures throughout the year. The difference between the warmest and coolest months may amount to only three or four degrees, but the wet season is characterized by high humidities.

The seasons of the subtropics are produced by the shifting of high-pressure cells, which move poleward in summer and back toward the equator in winter; variations in the Intertropical Convergence Zone, a belt of rainshowers and thunderstorms circling the equator; and monsoonal winds, which blow clouds and rain out to sea in winter. In some subtropical regions, the transition from tropical areas is marked by monsoon forests and cloud forests.

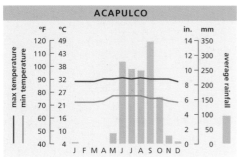

SEASONAL CYCLES
Acapulco, Mexico, has elevated temperature year-round, with a slight cooling in the northern hemisphere winter. Rainfall exhibits a pronounced dry season/wet season cycle, with the peak around June to September; winter rain is minimal.

Nesting in the wet Australia's Kakadu National Park boasts more than 200 bird species. As the wet season approaches, waterbirds such as magpie geese search the region for nesting sites.

Eating on the run The emerald hummingbird flourishes on the flowers of subtropical Central and South America. With its very high metabolism, the tiny bird must consume great quantities of nectar.

Aquatic refuge Northern Australia's Yellow Water Billabong —a waterhole in a branch of a river—is crowded with waterbirds when the surrounding floodplains dry out at the end of the wet season.

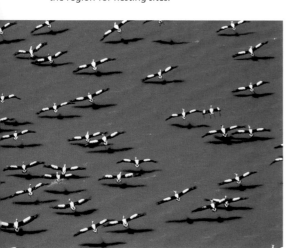

Misty mountains Along the slopes of mountains in Africa, Central and South America, Indonesia and New Guinea, the air cools and water vapor condenses into clouds, giving rise to "cloud" forests. The lower temperatures at high altitude allow for a mixture of temperate and tropical species.

■ WORLD SUBTROPICAL ZONES

Seasonal variations in the subtropical high-pressure cells, the Intertropical Convergence Zone and the Indian monsoonal regime create subtropical climates in much of South America, central Africa, southern and eastern Asia, and northern Australia. Temperatures are lower and rainfall more seasonal than in the tropics.

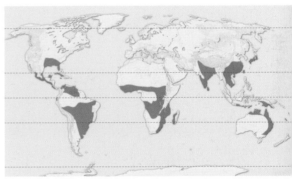

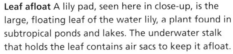

Weatherproof To minimize water loss during the dry season in the subtropical forests of Central and South America, the green iguana excretes salt from its nasal glands and has waterproof keratin on its skin.

Leaf afloat A lily pad, seen here in close-up, is the large, floating leaf of the water lily, a plant found in subtropical ponds and lakes. The underwater stalk that holds the leaf contains air sacs to keep it afloat.

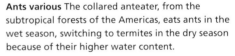

Ants various The collared anteater, from the subtropical forests of the Americas, eats ants in the wet season, switching to termites in the dry season because of their higher water content.

Arid

Arid climates typically create deserts, where the annual precipitation is less than 10 inches (250 mm), and high temperatures ensure that evaporation exceeds this precipitation. Deserts are subject to huge daily temperature fluctuations. The lack of cloud cover allows temperatures to soar during the daytime but fall rapidly after the Sun sets. Many arid zones lie under constantly sinking air such as near subtropical high-pressure cells or downwind of mountains, resulting in cloud-free sky and dry conditions. Some deserts occur in continental interiors, where little moisture arrives from the oceans; coastal deserts are located beside cold ocean currents that suppress precipitation. Despite minimal rainfall, high temperatures and drying winds, many plants and animals have adapted to these harsh environments.

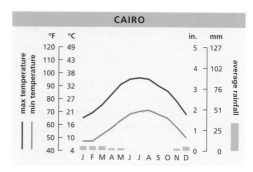

CAIRO

HOT AND DRY
Although rainfall tends to remain low year-round in Cairo, temperatures peak midyear when the Sun is high in a mostly cloud-free sky. Relatively low humidity in the region means large temperature swings between day and night.

↓ **Fleeting beauty** When rain does come in the desert, a profusion of colorful wildflowers may suddenly appear. These are ephemerals, annual plants that germinate only after rain and complete their life cycle within a few months.

↑ **Safe storage** Native to the Sonoran Desert of Mexico and the United States, the saguaro cactus stores water from sporadic rains in its stems and roots. Its waxy skin reduces water loss.

← **At the wind's mercy** Vast sand dunes occur in Death Valley in California, USA. Persistent winds pick up the sand and pile it into dunes, while swirling eddies sculpt the dune sides.

↓ **Ship of the desert** Camels survive without water for days by eating juicy plants; they minimize heat gain by storing fat in their humps instead of their bodies. Their broad, cloven hooves allow them to cross the hot sand without sinking. Their long nasal passages can be closed during sand storms.

WEATHERBEATEN

Arid landscapes, such as Monument Valley in Arizona and Utah, USA, are often barren with extensive exposures of bedrock. Lack of rain slows the forming of soil by weathering. Only about 25 percent of the world's deserts are composed of sand; most are rugged, rocky landscapes such as these. Indeed, while the popular perception of the Sahara, Earth's largest desert, is of a sandy wasteland, sand dunes cover only about a quarter of its landscape. The main subtropical arid zones of the northern hemisphere are the Sahara, Arabian and Thar deserts, and the deserts of the southwestern United States and Mexico. In the southern hemisphere the major subtropical deserts are the Namib and Kalahari of Africa, the Atacama in South America and the deserts of central Australia.

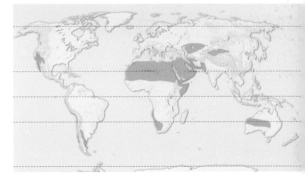

■ WORLD ARID ZONES

Most of the world's arid climate zones lie beneath the large subtropical high-pressure systems near 30°N and 30°S latitude. Major deserts occur in northern and southern Africa, southwest Asia and Australia.

Semiarid

The semiarid climate zones feature large expanses of grasslands and savannas because the annual precipitation ranges from 10 to 30 inches (250 to 760 mm)—enough water to support some vegetation but too little to sustain full forests. Semiarid regions extend from the tropics into the middle latitudes, wherever passing weather systems supply some moisture. Periods of severe drought also regularly occur. With few trees, these flat, exposed regions are very windy. The grasslands are maintained by sweeping seasonal fires and grazing by large herd animals, which clear the buildup of thatch and remove competing woody plants. Most of the grasslands' living mass lies underground. The thick roots of grasses store nutrients and collect moisture from the soil. After fire or grazing, the root system colonizes new areas.

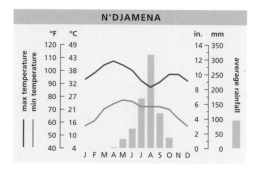

N'DJAMENA

°F / °C — max temperature / min temperature — in. / mm — average rainfall

J F M A M J J A S O N D

SUMMER RAIN
N'Djamena, in the Sahel area of Chad in Northern Africa, is hot throughout the year, but with slightly lower temperatures between July and September. This period corresponds with the wet season, when more than 80 percent of the year's rain occurs.

Grass survival The narrow, waxy leaves of grass plants conserve moisture, while deep, creeping roots store nutrients, collect moisture from the soil and allow quick recovery from fires and grazing animals.

Grass croppers Large herds of hooved grazers such as zebras and gazelles dominate the grasslands. Native to the semiarid plains and brush country of Africa, zebras spend up to 75 percent of their time grazing.

Semiarid display African daisies, hardy perennials with showy, daisy-like flowers, are native to the semiarid climate of South Africa. They need sunny conditions for the flowers to fully open.

All the way to the horizon The grasslands of Badlands National Park in South Dakota, USA, grow in the rain shadow cast by the Rocky Mountains. They are part of North America's vast prairies—a region of 1.4 million square miles (3.6 million sq km) covered with perennial grasses.

■ WORLD SEMIARID ZONES

Semiarid grasslands are referred to as veldt in Africa, prairie in western North America, steppe in southern Russia and pampas in South America. Particularly associated with Africa, savannas are grasslands with sparse tree growth that support herds of hooved grazing animals.

Prairie squirrel Prairie dogs are burrowing ground squirrels found across the Great Plains of North America. Thousands live together in "towns" of interconnected burrows beneath the prairies.

Sharp defenses The African savanna's acacia tree has thorny branches to discourage grazers. Small leaflets prevent water loss and a deep taproot absorbs soil moisture during the dry season.

The rain "shadow" Clouds release most of their precipitation on the windward side of mountains. As windborne air descends the lee side, it becomes warm and dry, resulting in a semiarid climate.

Mediterranean

A Mediterranean-type climate is characterized by warm, dry summers and mild, wet winters. It is created by seasonal variations in the position of subtropical high-pressure cells over western sections of the major continents. During summer, these cells drift poleward and their eastern flanks keep the regions dry and warm with sunny skies. In winter, however, the highs drift back toward the equator, permitting rain-bearing midlatitude storms to traverse the regions. Scrublands dominate in Mediterranean climates. This vegetation can survive the droughtlike conditions and wildfires that are a feature of the warm, dry summer months. Annuals, such as poppies, are also well adapted to this climate; their seeds remain dormant in summer and germinate with the winter rains. Mammals include deer, rabbits and rodents.

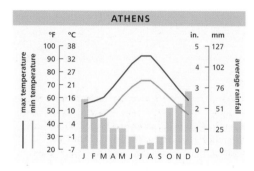

WET AND DRY
The Mediterranean pattern of hot, dry summers, followed by cool, wet winters, can be seen in this graph of the average monthly temperatures and rainfall experienced in Athens, Greece. Sea breezes help to moderate the heat.

↓ **Scrub of many names** Known as chaparral in California, Mediterranean scrub communities are called maquis around the Mediterranean Sea, mattoral in central Chile, fynbos in South Africa's Cape region, and mallee in Australia.

↑ **Clever jay** The scrub jay is omnivorous, allowing it to make the most of available food sources. During dry times, it can obtain all its water requirements from its food—a valuable asset in dry Mediterranean climates.

← **Food trees** Drought-resistant olive and almond trees, native to the Mediterranean, have been cultivated since ancient times as a food source. Olive trees need a cool winter to set the fruit that ripens in the summer sun.

↓ **Drought resistance** Eucalyptus trees and koalas are both adapted to Australia's Mediterranean climate. Eucalypts are succulents with leathery, grayish leaves. Koalas rarely need to drink—they obtain all the water they need from the foliage.

SUMMER PHENOMENON

During extended dry spells, wildfires can erupt throughout Mediterranean zones. Many plants that thrive in the regions, such as sagebrush, eucalyptus and rosemary, have aromatic foliage full of oils that add explosive fuel to the fires. Some species need fire to survive: they may lie dormant for years until fire germinates their stored seeds. Mediterranean climates favor shrubs over trees: their compact shape reduces moisture loss in the hot summers, and their extensive root systems gather scarce nutrients from the soil. Mediterranean shrubs often have thick, waxy leaves that help reduce water loss in hot, dry weather. Some have leaves with spiked margins that discourage grazing animals; some are covered with hairs to trap air and dissipate heat.

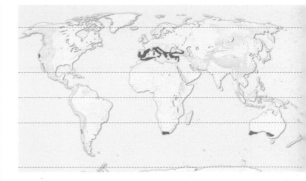

■ WORLD MEDITERRANEAN ZONES

Mediterranean climates mostly occur along coasts. Climates similar to those of the coastal areas around the Mediterranean Sea also occur in the United States, South Africa, Chile and southern Australia.

Temperate

Temperate climates are found in midlatitudes where nearly half the months have temperatures above 50°F (10°C). These regions all experience four distinct seasons, but the severity of the winter varies according to their proximity to the sea. Along the western edges of the continents, the prevailing ocean winds tend to create a temperate oceanic climate where the lowest monthly temperature rarely falls below 32°F (0°C), with essentially no winter snow cover. Elsewhere, temperate continental climates may have one or two months of persistent snow. Annual precipitation is generally adequate, though freezing winter temperatures can lock up moisture as snow and ice. The dominant vegetation is deciduous forest. Animals need to survive cold winters and seasonal variations in their food supply.

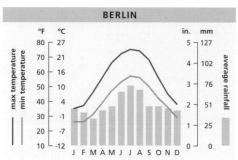

BERLIN

MILD AND EQUABLE
The German capital, Berlin, has cold and sometimes snowy winters, but warm summers in which the maximum temperatures typically reach about 75°F (24°C). Precipitation shows a summer peak but is relatively uniform throughout the year.

SEASONAL CHANGE

Spring As daylight lengthens, many plants form buds and leaves as migratory animals return.

Summer With sunlight near maximum intensity, plants store food as they display lush green growth.

Autumn The growing season draws to a close and deciduous trees change color as daylight decreases.

Winter By winter, many plants have become dormant, while animals have migrated or become inactive.

Fall shades When daylength diminishes and temperatures fall during the autumn months in a temperate zone, the chlorophyll that colors deciduous tree leaves green disappears, revealing the orange and red of the leaves' carotene and anthocyanin. The northeastern region of the United States is famous for its spectacular fall colors.

■ WORLD TEMPERATE ZONES

Temperate climate zones are found in the middle latitudes along the western coasts of North America and Europe, across eastern North America and eastern Asia, and in southeastern Australia and New Zealand.

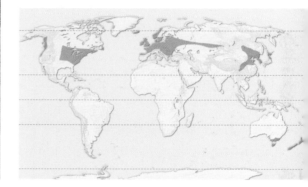

Putting something by Some temperate-zone animals remain active throughout the year. Squirrels store food in holes during late summer for use during the cold and snowy days of winter.

Lavender blue During spring and summer, Provence in the south of France displays magnificent fields of lavender, a herb used in perfumes and medicines. The summer sun brings out the essence of the flower.

Skeletal spinney Deciduous forests are the dominant vegetation in temperate regions. Before winter's onset, leaves wither and fall as protection from the cold and often droughtlike conditions.

Northern temperate

Since continental landmasses with extreme seasonal temperature contrasts are found primarily in the northern hemisphere, such regions are known as the northern temperate (or boreal) zone. With their large tracts of coniferous forest, they mark a transition between the Arctic tundra to the north and the temperate forest to the south. Typically, the northern temperate zone is marked by a relatively low annual average temperature, with strong seasonal contrasts provided by long, cold winters and short, cool summers. At least one month has an average temperature exceeding 50°F (10°C). Annual precipitation is small, with snow in winter and relatively heavy summer rainfall. The soils become waterlogged in the spring thaw and often remain soggy because of an underlying layer of permanently frozen soil.

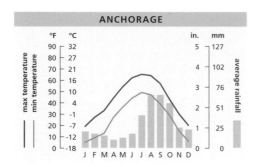

ANCHORAGE

HIGH CONTRAST
Anchorage, in Alaska, USA, experiences long and very cold winters followed by brief, cool summers. The overall climate is quite dry, but with a rainfall peak during late summer to early autumn months. Wind chill is a natural hazard here.

World of trees Northern temperate forests are dominated by conifers such as spruce which have a conical shape that allows snow to slide off, while their needlelike leaves withstand cold, dry winters.

White on white The snowshoe hare has large paws that permit it to travel over fresh snow cover. As winter approaches, the hare's brown summer coat turns white for better camouflage.

Leafy heaths Bunchberry and blueberry plants are members of the heath family, with succulent leaves that cut moisture loss in the relatively dry climate. Their roots absorb nutrients from the sandy soil.

How to stay warm Reindeer and caribou have thick coats that help them to survive the winter cold. Their large size limits heat loss, while broad hooves allow them to walk over the soft snow cover of winter and the boggy ground of summer.

■ WORLD NORTHERN TEMPERATE ZONES

The northern temperate climate stretches across subpolar North America and Eurasia, coinciding with the coniferous forests. The southern boundary of this climate is along the temperate zone at about 50–55°N latitude, while the northern boundary is marked by the timberline, near the Arctic Circle. Nine months of cold harsh weather are common.

Hardy heathers Heathers carpet the floor of this northern temperate forest; these low shrubs grow close together and can survive the snow cover of winter. Their small leaves reduce water loss.

Icy taiga The transition from northern temperate to polar climate zones tends to be marked by open forests of conifers, mixed with birch and aspen trees. The Russians call such forests *taiga*.

South for the winter In spring, snow geese migrate from southern climes to their nesting grounds in the northern temperate zone. Before winter arrives, they return south to their wintering grounds.

Polar

The polar regions of both the northern and southern hemispheres experience extended intervals of darkness and light throughout the year. With long winter darkness and relatively weak summer sunshine, the average temperature of the warmest month typically is less than 50°F (10°C). Precipitation, mostly snow, is also relatively light, with annual totals usually less than 10 inches (250 mm). This harsh climate results in a landscape covered by low-growing tundra vegetation or barren ice cap. The equatorward boundary of the polar zone is defined by the timberline, where the last hardy trees grow before the treeless plains of the tundra begin. Arctic plants include hardy grasses and sedges; animals include large warm-blooded wolves and bears. The Antarctic supports lichens and mosses, whales, seals and birds.

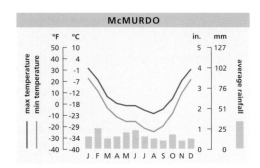

COLD AND DRY
Showing the classic profile of extremely cold winters and low precipitation year-round, the polar climate of Antarctica's McMurdo station is harsh. Only highly specialized lifeforms can survive these inhospitable conditions.

Antarctic emperor Emperor penguins are diving birds of the Antarctic coast and surrounding waters of the Southern Ocean. A thick layer of blubber and stiff waterproof feathers insulate them from the cold.

Bleak prospect The tundra is a cold arctic desert of the northern polar zone where the only plants are perennials, such as grasses and sedges, that survive a very short growing season with little precipitation.

Midnight Sun During the summer months in polar regions, the Sun remains just above the local horizon throughout the 24-hour day. Despite the extended daylight, snow and ice can remain year-round.

Icy immensity Northern Europe's largest glacier, the Jostedalsbreen, is a remnant of a large ice sheet that covered Norway as recently as 10,000 years ago. One of the glacier's arms, the Nigardsbreen (*left*), moves down a valley where the rugged surrounding landscape has been shaped by glacial erosion.

☐ WORLD POLAR ZONES
Polar climates are found poleward of the Arctic and Antarctic circles. In the northern hemisphere, the polar zone includes lands circling the Arctic Ocean. The entire Antarctic continent resides within the southern hemisphere's polar zone. These areas are dominated by snow and ice.

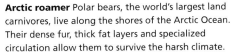

Tenacious plants Along with grasses, lichens are among the few plants that can survive in the subfreezing temperatures of Antarctica. They tend to grow on dark, heat-absorbing rocks.

Arctic roamer Polar bears, the world's largest land carnivores, live along the shores of the Arctic Ocean. Their dense fur, thick fat layers and specialized circulation allow them to survive the harsh climate.

Permanent winter Most precipitation in the polar climate zone is snow, which can remain throughout the year, forming glacial ice as it compacts. Pack ice surrounds the landmasses for much of the year.

Coastal

The climate along the shore of an ocean or other large body of water is considerably different from that found even a short distance inland. Because the temperature of the near-surface water changes quite slowly throughout the year, temperature variations along the coast are delayed, producing a relatively stable climate with few fluctuations in temperature. Sea breezes have a moderating effect, lowering temperatures on summer days. In spite of their stable temperatures, coastal regions can be harsh environments. Plants and animals have to survive strong onshore winds, waves, salt spray and windblown sand. Sandy shores and dunes can become desert-like as water percolates quickly through the sand. Rocky shores are exposed to pounding surf and constant soaking and drying.

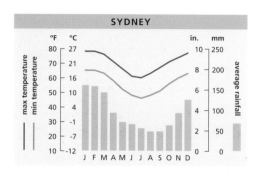

SYDNEY

MINOR VARIATIONS Sydney, Australia, has a generally mild climate—summer temperatures are moderated by a pronounced sea breeze and rainfall is fairly uniform. The driest period of the year is around the winter–spring months of August and September.

↓ **Weatherbeaten** The plants of Central America's Yucatán shoreline survive in a harsh climate. They must be drought-adapted because the sandy or rocky soil retains little moisture, and tend to be low-growing to withstand the winds.

↑ **Beach nursery** Like other sea turtles, the leatherback comes ashore only to lay and bury eggs on beaches. After the warm sand incubates the eggs, hatchlings must scramble across the beach to the ocean, dodging predators as they go.

← **Watery forests** The Mangrove Islands in Florida's Everglades National Park, USA, are formed by dense forests of mangroves that crowd the fringes of saltwater channels and freshwater rivers near the coastline.

↓ **Salt-tolerant trees** Mangroves grow in coastal environments, along muddy shorelines and in tidal estuaries and salt marshes. The trees limit dehydration by excreting salt through their waxy, succulent leaves.

Fish catcher The Atlantic puffin is a diving seabird that nests in the cliffs along the North Atlantic Ocean. Its bill has a serrated edge that allows it to carry several slender fish back to its young. The most obvious coastal habitats for birds are cliffs, rocks, sandy beaches, muddy estuaries and mangrove swamps. While there is some overlap, certain bird groups are associated with each habitat: terns and sanderlings with beaches, cormorants and puffins with cliffs, shorebirds with estuaries. Mangroves are especially attractive to long-legged wading birds such as ibises, spoonbills and egrets. Some coastal birds nest in burrows; some in dense colonies on the shore; others on cliff ledges (the eggs of these are narrower at one end than the other to prevent them from rolling off the ledge). Climate is a major factor in the habitat and type of food available to coastal birds.

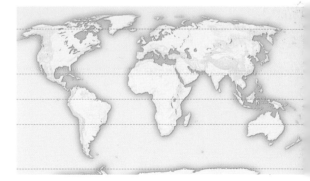

■ **WORLD COASTAL ZONES**
Coastal climates occur along all the world's ocean shores, regardless of latitude, but the differences between the coastal and inland climates tend to be more accentuated in middle to subpolar latitudes.

Weather and landform

Environmental agents such as wind, water and living organisms break down Earth's rocks through processes that are known as weathering. Some weathering processes are physical, involving frost action, water action, or changes in pressure or temperature. These break down the rock without altering its chemical composition. Chemical weathering occurs when water, sometimes carrying dissolved materials, alters the chemical composition of the rock minerals and causes it to crumble. In biological weathering, plants or animals disintegrate rock through either physical processes (such as the intrusion of roots or burrows) or chemical processes (such as the decay of vegetation).

The weathering process prepares the way for erosion, as the sediments it creates may then be removed by wind, water (rain, rivers, streams and oceans) or ice (snow and glaciers). As the sediments are transported, their abrasive action gradually sculpts the landscape.

→ **Natural sculpture gallery** At Cedar Breaks in Utah, USA, rock pillars known as hoodoos poke through the snow. The hoodoos were created when water that had seeped into cracks in the sandstone rock froze and expanded, gradually widening the cracks until columns formed. Different shapes emerged as layers of varying hardness eroded at different rates.

↓ **Growing old with polish** Bristlecone pines grow in the arid mountains of the American Southwest and are among the world's oldest living organisms Here, the high winds that force the trees to grow in twisted and contorted shapes also carry the ice and sand that polish their bark.

↓ **The shape of things** Daily temperature swings can lead to "exfoliation" in which outer layers of rock heat and expand more than the inner rock and are weakened, eventually crumbling to leave rounded boulders. Temperature variations can also detach blocks of rocks from their main slab and and "frost wedging" can split rocks through ice expansion. The constant forces of weather are a major contributor to the recycling of rocks—a continuous breakdown and rebirth, growth and decay.

EXFOLIATION

BLOCK DISINTEGRATION

→ **The power of water** Weathering of rocks produces debris that may end up in streams. As this sediment is carried downstream by the running water, it wears down and polishes exposed rock. The dislodged particles add to the stream's load of sediment, which will be deposited when the stream loses power or enters a lake or sea.

↓ **Frost-formed rubble** The summit of Glyder Fach Mountain, in Wales, is subject to repeated freeze–thaw cycles. Rainwater trickles into fractures in the rhyolite, then expands as it freezes, slowly prizing apart the rock. This process has produced jagged rock formations.

→ **The headstones of the Pinnacles** These rock pillars in the state of Western Australia were formed over thousands of years as slightly acidic winter rains dissolved relatively soft limestone formations, and winds blew away the overlying sand. The result is a startling array of outcrops amid the desert sand.

FROST WEDGING

Adaptations to weather: plants

Like animals, plants can be found in most places on Earth, even in environments that seem hostile to life. Although the available water, air temperature, light intensity and carbon dioxide concentrations vary dramatically around the globe, plants have developed characteristics that allow them to take advantage of the local conditions.

Most plants collect water through a root system, and gather the Sun's energy through leaves or needles, which have stomata (pores) that regulate water loss by opening and closing. In dry conditions, the stomata close to help the plant retain water. Drought-adapted plants may have deep roots to reach water far below the surface, or shallow, extensive roots to capture as much rainwater as possible. They are often succulents, with fleshy stems and leaves that store water.

In warm areas with distinct wet and dry seasons, or in cold climates where winter snow and ice lock up moisture, many plants are deciduous, losing their leaves during the water-scarce season. Outside tropical areas, many plants survive the subfreezing weather and reduced light intensity of the winter months by becoming dormant. Tundra plants such as the Arctic avens grow close to the ground to obtain warmth, and have thick hairy leaves to trap heat and prevent water loss. Hardy lichens, algae and mosses even find a niche in the bitter climate of Antarctica, having adapted to a scant supply of nutrients, almost no soil and limited water. Dormant for most of the year, they grow so slowly that their growth is measured in centuries.

↓ **Hardy rock colonizers** These rocks are home to lichens—small, slow-growing composites of fungi and algae that can survive in harsh environments. Attaching themselves to the surface with tiny hairlike growths, the lichens obtain all their nutrients from the surrounding air and bare rock.

→ **Trunkful of water** Found in semiarid regions of sub-Saharan Africa, baobab trees store water in their bottle-shaped trunks. They are deciduous, shedding their leaves during the long dry season, and producing new leaves and flowers when the annual wet season brings rain.

Relentless wind In windy areas, vegetation often grows in the direction dictated by the prevailing winds. Strong, persistent winds have bent these branches in the downwind direction.

Prickly plants Native to desert regions of the Americas, cactus plants are a group of succulents featuring a fleshy green stem covered in spines. They cope with the very dry climate by storing water in the fleshy stem, having a waxy surface, and growing few or no leaves. A shallow but extensive root system collects as much water as possible from dew or infrequent rains, before it disappears through the porous desert soil or sand. Their pleated surfaces enable these plants to expand or contract as they absorb or lose water, and the prickly spines deter grazing animals. Common succulents include the prickly pear, cholla and agave.

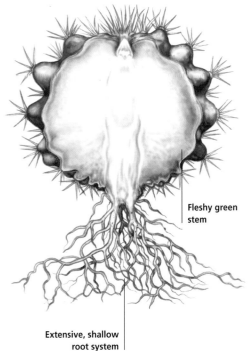

Fleshy green stem

Extensive, shallow root system

Individual reservoir Bromeliads grow in the tropical rain forests of Central and South America. They often grow on tree branches, gathering rainwater in the reservoir created by their overlapping leaves. To attract pollinating insects, the leaves turn a brilliant red for the duration of the bloom period.

Coastal reinforcement Mangroves grow in dense thickets in tropical salt marshes and tidal estuaries. Their exposed roots hold the trunk above the water line, and carry oxygen to tap roots below the mud. The tangled roots reduce coastal erosion by capturing sediment from rivers and streams before it enters the ocean.

Adaptations to weather: animals

Over millions of years, animals have evolved to survive in Earth's many different climates, developing adaptations that allow them to cope with the weather. These specializations may be physical, such as an insulating layer of blubber, or behavioral, such as a burrowing habit. Many animals simply avoid temperatures that are outside their optimal range. Some hibernate or become torpid; others migrate whenever the weather in their habitat turns too cold or too hot.

Birds and mammals are endothermic (warm-blooded), maintaining a constant core temperature by processing body fat and carbohydrates. To stay active through cold winters, endotherms must eat constantly. Layers of feathers, fur or fat help the body retain heat in cold climates, while evaporative cooling methods, such as sweating or panting, help in warm conditions. All other animals, including fish and reptiles, are ectothermic (cold-blooded), with their body temperature governed by the temperature of the environment. To adjust their core temperature, ectotherms must move into or out of sunlight or water.

While biodiversity is greatest in the warm, wet tropics, animals have also adapted to seemingly hostile conditions in the polar regions, arid deserts and mountaintops. Although the number of species is lower in high latitudes, this is partially compensated for by species density.

Arctic hare | Snowshoe hare | Black-tailed jackrabbit | Antelope jackrabbit

↑ **At home in the snow** Polar bears have a blood adaptation that reduces heat loss. Warm blood from the heart transfers heat to cool blood returning from the skin. Thick fur, which extends to the soles of the feet, provides perfect insulation. A layer of blubber under the skin not only insulates the bears from the cold but also retains body heat when they swim under the icy sea.

← **Long and short** An endothermic (warm-blooded) animal's size and shape affect the amount of body heat it loses, with a larger surface area increasing the heat loss. In contrast to hares and rabbits in warmer climates, Arctic and snowshoe hares have shorter ears and legs, thereby minimizing their body surface area and helping them survive cold winters. In the subtropics, the black-tailed and antelope jackrabbits have longer limbs and ears, which help dissipate body heat on warm days. Their toes are widespread to enable them to travel easily over soft snow, and their fur changes from summer brown to winter white to provide year-round camouflage.

← **Camouflage coat** Arctic foxes are well adapted to polar climates and stay active through winter. Their thick coats—70 percent of which are warm underfur—reduce heat loss and provide year-round camouflage. Most foxes have a thicker, whiter coat in winter; this thins and turns brown in summer. A small percentage of Arctic foxes, usually those near the ice-free oceans, have a bluish gray coat that thins and turns chocolate brown in summer.

BEATING THE HEAT

Deserts are not nearly as deserted as their name implies; indeed they often support an abundance of animals. The main threats to desert animals are excessive heat and dehydration, and animals must maintain their body temperatures at tolerable levels while restricting water loss. Adaptations include an ability to eliminate waste products in a minimum amount of fluid, the growth of large sacs for storing food, burrowing, becoming dormant as temperatures soar and nocturnal hunting. This Peringuey's sidewinding adder (*above*) waits for prey camouflaged in baking sand in the Namib Desert.

←← **Relaxed attitude** Sloths spend most of their time in tropical rain-forest trees, sleeping or eating. To minimize heat production, they remain sluggish with a very slow metabolism. In the high humidity, algae grows on their fur and provides camouflage from predators such as the harpy eagle and jaguar. Sloths descend to the ground only to defecate and move clumsily along the forest floor.

← **Burrowing away** Found throughout much of sub-Saharan Africa, the nocturnal aardvark avoids daytime heat by sleeping in its underground burrow, emerging at night to forage for ants and termites. Their large ears act as radiators to assist in temperature regulation, as with other desert dwellers such as the jackrabbit.

Adaptations to weather: humans

Humans are endotherms (warm-blooded animals) who need to maintain a constant core temperature near 98.6°F (37°C). Variations much above this temperature can lead to dehydration and the potentially fatal condition of hyperthermia; variations much below can produce frostbite and hypothermia, a progressive physical and mental collapse.

In a warm environment, the human body dissipates heat by increasing blood flow to the extremities. Especially warm conditions or physical activity will trigger sweating, in which the skin is cooled as perspiration evaporates. It takes only a week or so for people to acclimatize to moderate heat, as their sweating and circulatory mechanisms become more efficient.

In a cold environment, the human body initially conserves heat by constricting the blood vessels underneath the skin. This is often followed by shivering, which generates extra heat by increasing the body's metabolic rate. Humans have a low tolerance of cold, however, and are generally unable to acclimatize. They therefore depend on the protected microclimate provided by layers of clothing, warm shelters and artificial heating. Throughout human history, the focus of much scientific endeavor has been to find ways of enabling humans to live more comfortably in their environment.

↓ **Sweating it off** Physical activity increases the body's metabolic rate and causes an increased output of sweat. As sweat evaporates, skin temperature is lowered. In extreme conditions, the body may lose up to 4 quarts (4 l) of fluid per hour.

→ **Dressed for the weather** An Inuit from the Nunavut territory in Canada hunts for fish in the traditional manner, using a harpoon and protected from the extreme cold in clothing made from caribou skins.

COOLING GLAND

Sweat glands are located in the subcutaneous tissue under the skin. They secrete watery perspiration through sweat ducts to the skin's surface. As the sweat evaporates, the skin is cooled. In humid conditions, perspiration is slow to evaporate.

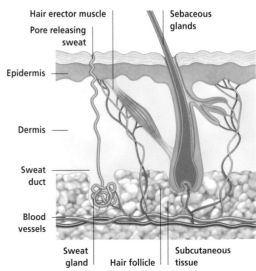

Hair erector muscle
Pore releasing sweat
Sebaceous glands
Epidermis
Dermis
Sweat duct
Blood vessels
Sweat gland
Hair follicle
Subcutaneous tissue

↑ **Roof efficiency** In snowy countries such as Austria, traditional architecture features steeply sloped roofs that allow much of the snowfall to slide off rather than accumulating and overloading the roof.

↑ **Natural insulation** In Tunisia's hot, dry climate, dwellings are dug into the ground to protect people from the intense daytime sunlight. Being slow to heat up and cool down, the ground acts as insulation, smoothing out the fluctuation from the extreme heat of day to the cold of night.

← **High living** At an elevation of 11,300 feet (3440 m), the amount of oxygen in the atmosphere of Namche Bazaar, Nepal, is only two-thirds the amount near sea level. To cope with this, visitors have to breathe more rapidly. Some people who live at high altitude have an increased amount of oxygen-carrying hemoglobin in the blood, allowing them to extract the required amount of oxygen from the air. Adapting to altitude also involves dealing with lower temperatures and drier air.

Harnessing the weather

For thousands of years, people have harnessed the Sun, wind and water, using them to generate power for heating, transportation, agriculture and industry. The nineteenth and twentieth centuries saw a massive shift to the fossil fuels of oil, coal and natural gas, but supplies of these resources are dwindling and concern has grown about the pollution they generate. In recent years, interest in renewable energy sources has revived, leading to the development of solar, wind and hydroelectric power technologies.

These renewable energy sources are dependent on the weather. Solar heating and electricity generation require relatively clear skies with bright sunshine. Wind energy needs reliable prevailing winds of sufficient strength to run mechanical or electrical power generators. Hydroelectric generators can work only if storm systems deposit sufficient rain or snow into the watershed upstream of the dam.

While renewable sources of energy are less polluting than coal-burning power stations, they do impact on the environment. Large areas of land are flooded to create the dams needed to provide water for hydroelectric schemes, altering ecosystems and displacing communities. Wind generators are unsightly and can be a hazard to birds. The effect of arrays of solar panels on desert climates is being monitored. Despite these drawbacks, however, the value of harnessing wind, water and sunlight is very real, when measured against the predicted effects of global warming.

↓ **Early windmills** These windmills near Campo de Criptana in Spain were built in the sixteenth century. The external rotor sails are attached to a main vertical shaft in the tower of each mill. As this shaft rotates, wooden cogwheels at the base translate the motion to turn millstones for grinding flour. Windmills also powered the bellows of iron forges.

↑ **Old-fashioned water energy** This Manangi woman is grinding grain on a water mill in the Annapurna region of Nepal, a practice that dates back centuries. The energy of a flowing river turns the blades of a water wheel, the movement of the wheel causes a millstone to turn, and this is used to grind grain. The ancient Egyptians were the first to use the power of water to turn stones for grinding grain; today they use water from the Aswan Dam to generate much of their electricity. Water still provides the most widely used form of renewable energy: hydroelectric power drives turbines that produce electricity. Prolonged drought can restrict hydroelectric capacity.

↑ **Harnessing the Sun's rays** As the power drops to zero at sunset, a technician prepares to wash the reflectors at the LUZ Solar Thermal Electric Generator at Kramer Junction, California, USA. By day, sunlight is concentrated by the curved mirrors, heating a special molten salt that circulates through the structure. As the hot, molten salt passes through heat exchangers, it creates superheated steam. The steam, in turn, drives a turbine that produces electricity. This is the world's largest solar power plant, generating sufficient power to supply a city of 350,000.

PUTTING WIND TO WORK

The turbines on this California, USA, wind farm are powered by rows of propellers that harness the wind—a non-polluting method of producing electricity. The greatest obstacle to the commercial development of wind power is the variability of wind strength, resulting in a similarly variable electrical output from wind turbines. A large-scale wind power system must therefore include mechanisms by which the energy that is generated in gusty periods can be stored for use when the wind is light. It is estimated that wind now supplies about one percent of California's energy. In addition to the large-scale wind farms shown above, wind power can be used to generate power at a domestic level.

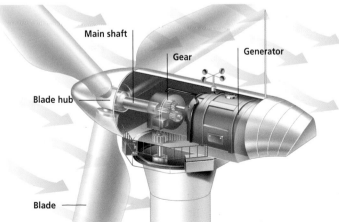

Main shaft

Gear

Generator

Blade hub

Blade

← **Contemporary model** Wind is, in fact, a form of solar energy as it derives primarily from air temperature differentials induced by the Sun (along with the shape and rotation of Earth). A modern wind turbine is an electrical generator mounted on a tower. The wind causes the propeller blades to turn, rotating a central shaft that is connected to a generator. The generator then feeds electricity to the utility grid for distribution to the consumer.

Modifying the weather

For thousands of years humans have attempted to exercise some control over the impact that weather has on their lives. The earliest measures took the form of prayers and ceremonies, such as the famous rain dance of North America's Hopi Indians. At a more practical level, early farmers developed agricultural practices such as irrigating crops in dry weather, protecting tender plants from frost, or planting shelterbelts of trees to reduce wind damage. Such practices continue today, but often use more sophisticated techniques and equipment.

With advances in the understanding of cloud physics during the twentieth century, scientists began intentionally seeding clouds to stimulate precipitation, clear fog, reduce hail size or weaken the force of hurricanes. While some of these cloud-seeding efforts have been moderately successful, they remain controversial since they are difficult to assess.

↓ **Watering the desert** In this satellite image, the green circles represent crops which are irrigated from large rotating booms. The water is tapped from fossil reserves deep underground.

→ **Anti-frost cover up** Night-time temperatures that fall below freezing can damage or kill tender plants. Here, a potato crop has been protected from frost by plastic sheeting.

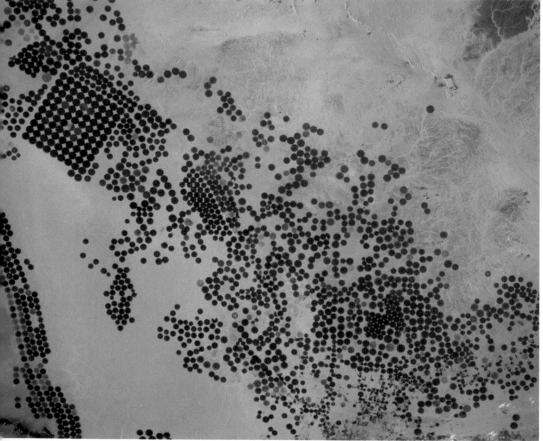

↑ **A cooling greenhouse** At this covered desert farm at Wafra on the border of Kuwait and Saudi Arabia, the crops are protected from the withering effects of direct sunlight. Environmentally controlled greenhouses can shade crops and provide cooled air to reduce the rate at which moisture is evaporated. Humankind can raise crops in deserts only by managing that vital but scarce resource: water. The simple techniques developed over centuries to conserve and utilize water are still in use in many places, and new technology now provides additional ways to increase and use the water supply.

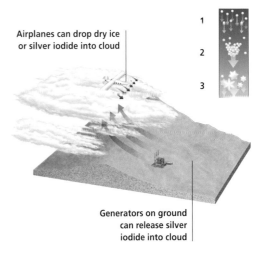

Airplanes can drop dry ice
or silver iodide into cloud

1
2
3

Generators on ground
can release silver
iodide into cloud

SEEDING THE CLOUDS

First employed in the 1940s, cloud seeding remains
a controversial method of modifying the weather.
It is used when the water droplets or ice crystals in
a cloud are not large enough to fall to the ground as
rain or snow. Generators on the ground or airplanes
in the sky send flares or fireworks-like devices to
release dry ice or silver iodide into the cloud.

1. Water droplets A cloud may be made up of tiny
droplets that are too small or unable to combine
into larger drops that can fall to the ground as rain.

2. Seeding agent The cloud may be seeded with
dry ice or silver iodide. Dry ice freezes the cloud's
tiny droplets into large ice crystals. With a crystal
structure similar to that of natural ice, silver iodide
also encourages the development of large crystals.

3. Ice crystals The large ice crystals created by the
seeding grow into snowflakes and eventually melt in
warm air to form raindrops.

← **Wearing the symbols of rain** This native American
boy at a traditional tribal ceremony has black vertical
lines painted on his face to symbolize the falling rain;
hail is sometimes represented by white circles.

Climate: inner planets

The climate of other planets depends primarily upon how distant the planet is from the Sun—planets close to the Sun will be warmer than more distant planets. The orientation of a planet's spin axis with respect to its orbital plane, and the planetary rotation rate, also influence its weather and climate. A planet with an axis perpendicular to the orbital plane will have fewer seasonal changes than one with a highly inclined axis. The faster a planet spins, the greater the range in its daily temperature variations. The composition of the atmosphere—which depends upon the planet's mass and distance from the Sun and is depicted here as cylindrical graphs—also plays a role in determining climate. This spread focuses on the climate of the planets closest to the Sun, the small and rocky terrestrial planets with relatively thin atmospheres.

MARTIAN SEASONS

Mars has a tilt similar to Earth's and experiences four seasons, each lasting about twice as long as Earth's because Mars' orbit is so much larger than Earth's. In the northern hemisphere, spring and summer generally feature a clear atmosphere with little dust in the air. Whitish clouds appear near the sunrise line and over high elevations. In autumn and winter, the northern polar cap vanishes under cloud.

→ **Clouds on Mars** In this image, taken from the Mars Global Surveyor Orbiter, several Martian volcanoes are visible in an area called the Tharsis region. The whitish areas of haze are water-ice clouds, possibly similar to the cirrus clouds seen on Earth.

VENUS STATISTICS
Discovered Known since antiquity
Farthest from Sun 67.7 million miles (108.9 million km)
Closest to Sun 66.8 million miles (107.5 million km)
Mean surface temperature 900°F (480°C)
Sunlight strength 190% of Earth's

- Other 0.8%
- Nitrogen 3.2%
- Carbon dioxide 96%

↓ **The greenhouse effect on Venus** Strong sunlight filters through the clouds and heats the surface, but the clouds and the carbon dioxide in the atmosphere prevent the heat from escaping. As a result, it is as hot at Venus' poles as at the equator, and the night side is no cooler than the day side. After filtering through the clouds, the light is colored orange and appears to be about as bright as an overcast day on Earth.

↑ **Sunrise on Mars** The Sojourner rover captured this image in July 1997. Mars, fourth planet from the Sun, may once have had oceans, or at least seas, of liquid water. Satellite views show apparent gullies that may have been carved by water (or perhaps carbon dioxide). Like Earth, Mars experiences dust storms, when dust-laden winds obscure the lava flows and boulder fields that make up much of its surface. Occasionally global dust storms obscure most of the planet's features.

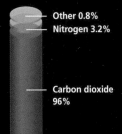

Incoming solar radiation

Cloud cover surrounding Venus

Most radiation is reflected back into space from the cloud tops

Some radiation passes through the cloud and heats the surface. This is then trapped within the atmosphere because of its high carbon-dioxide content

MARS STATISTICS
Discovered Known since antiquity
Farthest from Sun 154.8 million miles (249.1 million km)
Closest to Sun 128.4 million miles (206.7 million km)
Mean surface temperature 10°F (-23°C)
Sunlight strength 36–52% of Earth's

- Other 0.7%
- Argon 1.6%
- Nitrogen 2.7%
- Carbon dioxide 95%

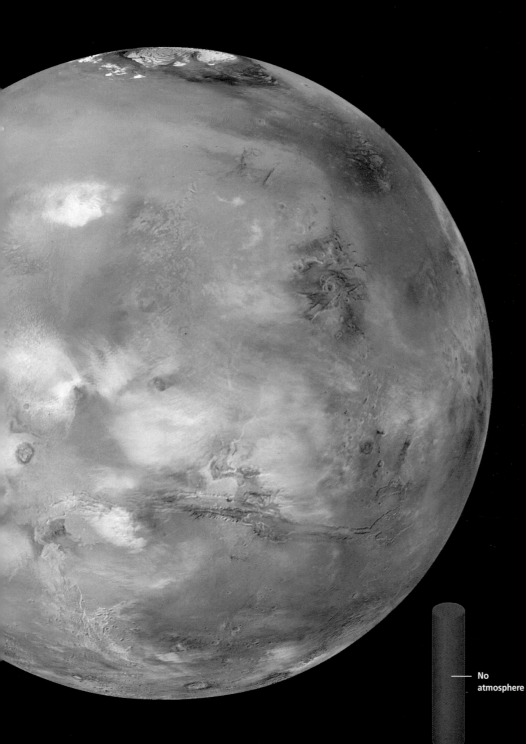

SURFACE TEMPERATURE OF MERCURY

Mercury's cratered surface *(below)* is the result of ancient impacts from meteorites and comets. It is the closest planet to the Sun and has the greatest range of surface temperatures. When the Sun is high overhead, the surface reaches around 800°F (430°C), but in the opposite hemisphere, with the Sun well out of view, it is more than 1080°F (600°C) cooler.

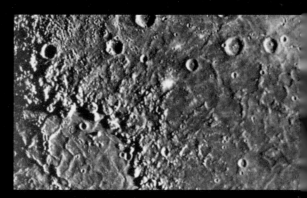

— No atmosphere

MERCURY STATISTICS
Discovered Known since antiquity
Farthest from Sun 43.4 million miles (69.8 million km)
Closest to Sun 28.6 million miles (46 million km)
Mean surface temperature 800°F (430°C)
Sunlight strength 450–1040% of Earth's

Climate: outer planets

The large, cold planets beyond Mars—Jupiter, Saturn, Uranus and Neptune—are known as the jovian planets or gas giants. Unlike Earth, they do not have a terrestrial mass inside their atmosphere; they are composed of lightweight gases and liquids, such as hydrogen and helium, rather than rocks and metal. And they are indeed giants: Jupiter is so large that more than a thousand Earths could fit inside it. The gas giants all have rings, the most famous of which are those of Saturn. Pluto is neither a rocky planet nor a gas giant—its interior is something of a mystery but it resembles the rock-ice moons of the outer planets. Astronomers today believe that Pluto wandered in to the Solar System from the Kuiper Belt, a region beyond the zone of the planets. The bodies in this region are called icy planetisimals, or comets without tails.

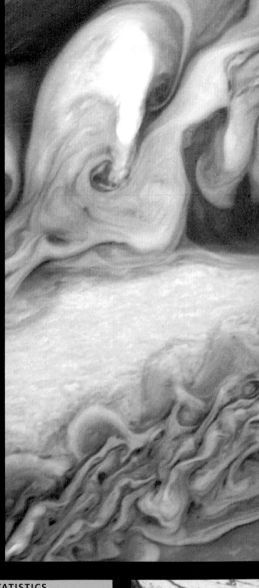

↑ **Hubble's Jupiter**
The Hubble Space telescope photographed this image of Jupiter in 1992. Jets of wind circulating in alternating directions create the bright and dark streams. Wild winds, blowing at 400 miles per hour (640 km/h), may be driven by convection from an internal energy source: Jupiter radiates about twice as much energy as it receives from the Sun.

← **Encircled planet**
The distinctive rings of Saturn are composed of thousands of orbiting ice particles and pieces of rocky debris produced by meteorite showers raining down on the planet. Blue indicates clouds.

JUPITER STATISTICS

Discovered Known since antiquity
Farthest from Sun 506.9 million miles (815.7 million km)
Closest to Sun 460.4 million miles (740.9 million km)
Mean surface temperature 240°F (-150°C)
Sunlight strength 3–4% of Earth's

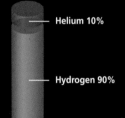

Helium 10%

Hydrogen 90%

Helium 4%

Hydrogen 96%

SATURN STATISTICS

Discovered Known since antiquity
Farthest from Sun 933.9 million miles (1503 million km))
Closest to Sun 837.6 million miles (1348 million km)
Mean surface temperature 110°F (-80°C)
Sunlight strength 1% of Earth's

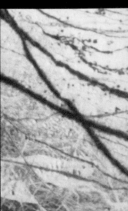

← **Neptunian cavalcade** Neptune is the smallest and most distant of the gas giants, but it still has a mass equal to 17 Earths. An ever-changing cavalcade of atmospheric bands, bright ammonia-ice clouds and dark stormy regions march around its bluish disk. Energy from within the planet powers violent winds and storms. The most famous of these, the Great Dark Spot, has since dissipated.

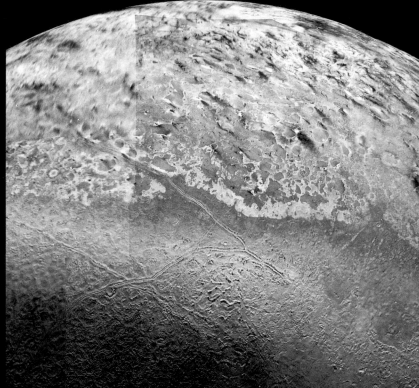

↑ **Jupiter's giant vortex** A closeup of the Great Red Spot on Jupiter, shown at top right of the image, indicates an intricate pattern of vortices in the surrounding area. The Great Red Spot, a spinning vortex more than twice Earth's size that has existed for centuries, appears to be similar to a large hurricane on Earth.

↗ **Neptune's moon** Triton is the largest of Neptune's 13 known moons. With a bitterly cold surface temperature, most of its atmospheric nitrogen is in the form of ice.

← **Jupiter's moon** Europa is one of Jupiter's many moons. Its icy surface resembles a planet-wide skating rink. Its surface is a shell of water-ice.

NEPTUNE STATISTICS

Discovered September 1846, Johann Galle
Farthest from Sun 2825 million miles (4546 million km)
Closest to Sun 2769 million miles (4456 million km)
Mean surface temperature 265°F (-220°C)
Sunlight strength 0.1% of Earth's

Methane 3%
Helium 18%
Hydrogen 79%

The changing climate

Earth's climate has changed throughout the history of our planet. Many of these fluctuations are short-term patterns, but movements of air, water and landmasses all contribute to long-term change. The human impact, too, is significant: when we alter Earth's ecosystem, the effects can be dramatic.

The big picture

Earth formed about 4600 million years ago, but there is little evidence of climate change for some 90 percent of our planet's lifetime. During the long, lifeless period before the appearance of algae—the first primitive forms of life—around 3500 million years ago, gases produced by volcanic eruptions probably accumulated to form the atmosphere and oceans that enable life to exist on Earth. Vast amounts of volcanic dust were thrown into the air and reflected away energy from the Sun, allowing the planet to cool: another contribution to an environment conducive to life on Earth. At some time between 2700 and 1800 million years ago, glaciers and ice sheets were widespread. Since then, Earth's climate has warmed and cooled over long periods of time.

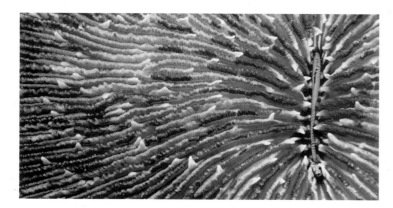

↑ **Coral clues** Coral colonies growing near the coasts are affected by river discharges that, in turn, are linked to rainfall. By examining cross-sections of coral under ultraviolet light, scientists can build up a history of tropical climate.

→ **Record of the rocks** The Bungle Bungle Range in Purnululu National Park, Western Australia, was deposited 375 to 350 million years ago. Ancient landforms such as this provide evidence of long-term climate change.

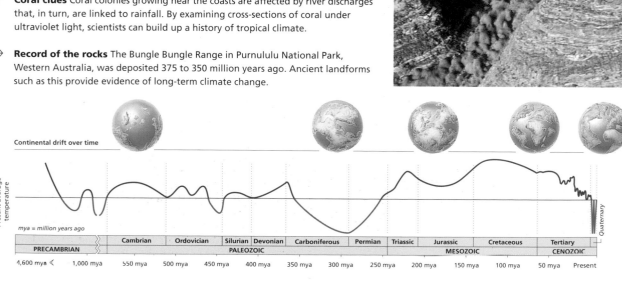

Tracking long-term climate The horizontal line shows average present temperature. The globes above the graph show the progression of continental drift. The long glaciation from 330 to 245 million years ago coincided with the formation of the supercontinent Pangea.

Continental drift over time

Present average temperature

mya = million years ago

	Cambrian	Ordovician	Silurian	Devonian	Carboniferous	Permian	Triassic	Jurassic	Cretaceous	Tertiary	
PRECAMBRIAN	PALEOZOIC						MESOZOIC			CENOZOIC	Quaternary

4,600 mya | 1,000 mya | 550 mya | 500 mya | 450 mya | 400 mya | 350 mya | 300 mya | 250 mya | 200 mya | 150 mya | 100 mya | 50 mya | Present

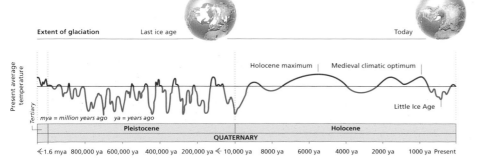

Extent of glaciation Last ice age Today

Present average temperature

mya = million years ago   ya = years ago

Holocene maximum   Medieval climatic optimum

Little Ice Age

Tertiary

Pleistocene	Holocene

QUATERNARY

1.6 mya 800,000 ya 600,000 ya 400,000 ya 200,000 ya 10,000 ya 8000 ya 6000 ya 4000 ya 2000 ya 1000 ya Present

CLIMATE CHANGE: AN ONGOING PROCESS

The Cenozoic Era covers the last 65 million years of Earth's history. For most of this long era, Earth's climate cooled, although this was no by no means a smooth and steady process. The second period of the Cenozoic is the Quaternary. The graph at left plots changes in temperature during this period, which began about 1.6 million years ago and includes the present time. In the first stage, the Pleistocene Epoch, there were seven major glaciations, with more than 30 percent of Earth covered by ice. We are living in the Holocene Epoch, a warm time that began some 10,000 years ago. We may well be living in an interglacial period that will end with yet another ice age. Despite all our scientific endeavor, our climatic future remains undetermined.

Long-term climate change

Long before people became concerned about the enhanced greenhouse effect on today's changing climate, scientists were aware that there had been a number of climate changes in Earth's 4600 million-year history. These long-term changes were caused by a complex web of forces, including movements of air, water, landmasses and Earth itself. The Sun provides the energy that drives our weather systems, so changes in incoming solar energy had marked effects on the Earth's climate. There are numerous reasons why the solar energy received at Earth's surface might change. These include the direct effects of meteorites and comets on our atmosphere, and variations in the energy output of the Sun itself. A large meteorite collision had the potential to produce a huge dust cloud that could circle Earth, block the Sun and produce an extended cool period. Collision with a comet, or even a close encounter, could produce a marked difference in the amount of solar energy reaching Earth's atmosphere. Geological evidence indicates that there have been five mass extinctions of flora and fauna, which may have occurred because of great climate changes.

THE SUN BECOMES BRIGHTER . . . OVER TIME

Scientific evidence points to a gradual increase in solar luminosity of around 20 to 30 percent since Earth formed about 4600 million years ago. This luminosity increase occurs as hydrogen in the Sun's core is converted to helium. Lesser variations are closely related to the periodic change in sunspot activity—the well-known 11-year sunspot cycle. Solar output will typically change by around 2 percent from a time of minimum sunspot activity to the corresponding sunspot maximum. There are also cycles within cycles; the very high to very low activity cycle takes about 80 years. Magnetic field fluctuations triggered by sunspot activity have a cycle of approximately 22 years, while lunar tide patterns recur every 19 years.

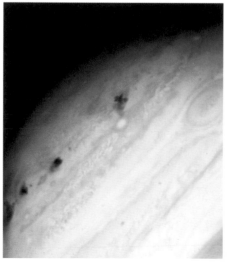

←← **The unprotected Moon** Unlike Earth, the Moon has no atmosphere to protect it from the impact of meteorites. When a piece of rock or ice hits the Moon with force, it blows up in a cloud of vapor and blasts a scar on the Moon's surface. Impacts range from small bowl-shaped craters to massive basins. The Moon is pitted by thousands of craters, evidence of the huge number of meteoroids floating in space that are also capable of crashing into Earth.

← **The bombardment of Jupiter** In July 1994, the Hubble Space Telescope captured these graphic, color-enhanced images of the huge, multiple craters caused by the impact of large fragments from the Shoemaker–Levy 9 Comet on the surface of Jupiter. The collision produced energy greater than every nuclear weapon on Earth, providing a graphic demonstration of what could happen if such an object hit our planet.

↑ **Prehistoric insult** The large meteorite that caused this snow-dusted crater in the Arizona desert crashed to Earth about 50,000 years ago. Widely scattered locations across the planet are pockmarked with the tell-tale craters that are a clear demonstration of the force and frequency with which meteorites plunge to the surface. On any dark, clear night, four or five meteors can be seen each hour; a meteor shower can produce a downpour of 20 to 50 meteors. A huge meteorite blast on Earth could create cooler conditions.

↗ **Shower from a cloudless sky** There have been several occasions when Earth has sustained major meteorite impacts, but most of the matter that is continually bombarding our planet's surface has little effect. This spectacular meteor shower is seen against the night sky of the California, USA, desert.

← **Silver streak** Comets are regular visitors to our Solar System. Some have visited only once in recorded history while others come into Earth's orbit at regular intervals. Impacting comets and their dust may have provided the gases that are the building blocks for life on Earth, but evidence also suggests that a comet or meteorite impact some 65 million years ago caused the mass extinction that claimed 80 percent of Earth's species.

Global forces at work

Of all the forces affecting long-term climate change, the most influential are the Sun and the oceans. The amount of solar radiation reaching the surface of Earth can change for a number of reasons. In the 1930s, a Yugoslav geologist, Milutin Milankovitch (*left*), discovered that periodic variations in the way Earth orbits around the Sun alter exposure to solar radiation. Analysis of deep-ocean sediments has shown evidence of these cycles over repeating periods of 19,000, 23,000, 100,000 and 433,000 years. Changes in ocean circulation patterns can lead to climatic shifts that may last from a few years to millennia. Volcanic eruptions spew dust and gas into the atmosphere, and possibly contributed to the ice ages. Even the inexorable process of continental drift, with its moving landmasses, has altered Earth's climate.

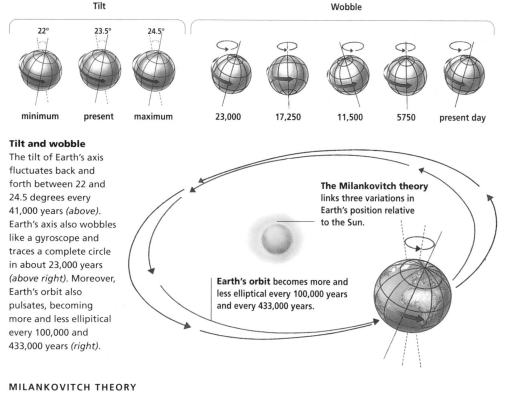

Tilt and wobble
The tilt of Earth's axis fluctuates back and forth between 22 and 24.5 degrees every 41,000 years (*above*). Earth's axis also wobbles like a gyroscope and traces a complete circle in about 23,000 years (*above right*). Moreover, Earth's orbit also pulsates, becoming more and less ellipitical every 100,000 and 433,000 years (*right*).

The Milankovitch theory links three variations in Earth's position relative to the Sun.

Earth's orbit becomes more and less elliptical every 100,000 years and every 433,000 years.

MILANKOVITCH THEORY

The Milankovitch theory links fluctuations of the ice ages with three variations in Earth's position relative to the Sun. The three periodic changes in Earth's path around the Sun are illustrated in the diagrams above. These variations cause radiation changes of up to 15 percent at high latitudes—changes that influence the expansion and melting of polar ice sheets. Although Milankovitch's hypothesis was greeted skeptically when first postulated, an increasing amount of scientific evidence now supports it.

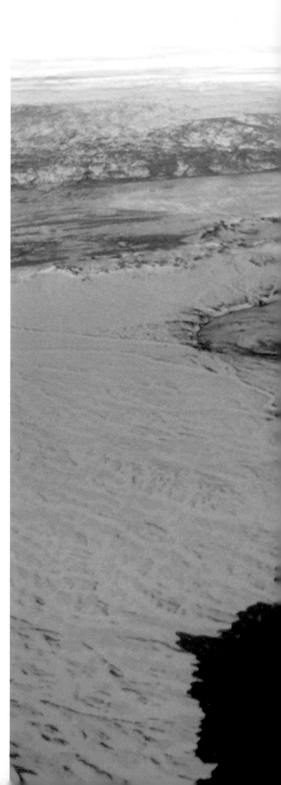

THE GREAT OCEAN CONVEYOR BELT

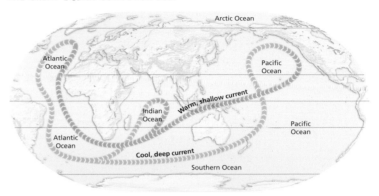

↑ **Round trip** The great ocean conveyor belt is a key factor in long-term changes in the surface temperature of the oceans. Cold, deep water from the Atlantic flows south and east, to resurface and be warmed in the Indian and north Pacific oceans. Currents carry warmer water back through the Pacific and south Atlantic. The round trip takes between 500 and 2000 years.

↑ **Clues in the ice** Scientists examine a 30-foot (10 m) ice core drilled from the Antarctic. Chemical analysis of ice cores can reveal changes in climate over several hundred years. Air trapped in the upper layer of frozen snow is untainted by the atmospheric air, and thus provides evidence of what the composition of the atmosphere was at the time when the snow froze. The three shorter cycles of climate change proposed by Milankovitch (19,000, 23,000 and 41,000 years) are clearly revealed in longer term studies.

← **Evidence from Iceland** Sun shines over the Grimsvoten Glacier, Iceland. The Milankovitch theory is supported by studies of sunlight falling in high latitudes .

Before the dinosaurs

Unraveling the mysteries of ancient climates requires extensive detective work. Only the oldest fossils and meteorite impact craters are able to provide the clues necessary to deduce what conditions were like in the millennia before dinosaurs walked on Earth. Knowing that each species of plant and animal has a particular range of climatic conditions in which it thrives, scientists are able to "read" fossil remains. By comparing the remains, geographical location and relative abundance of creatures that existed many millions of years ago with their modern counterparts, they can make deductions about past climate conditions.

The first sedimentary rocks were laid down some 3700 million years ago, when the climate was probably 18°F (10°C) warmer than today. Algae, the first lifeforms, appeared about 3500 million years ago. From then until the Age of Dinosaurs (250 million to 65 million years ago), Earth successively warmed and cooled, with periods of intense glaciation followed by more equable conditions.

PRE-DINOSAUR EARTH: A WARM, STEAMY PLACE

Much of the available fossil evidence points to the fact that during the pre-dinosaur period, Earth experienced an extended period—many millions of years—of warm and humid weather, with average temperatures several degrees higher than today. These warmer temperatures also affected the developing oceans, and rainfall over many of the existing landmasses was also probably higher than today. Plant fossils that have been preserved in rock for millions of years indicate that much of the vegetation consisted of varieties of ferns, which require warm, humid conditions to thrive.

↑ **Preserved in stone** As ferns need abundant moisture and high humidity levels to thrive, this specimen of a *Pecoptens* fossil fern from the Carboniferous period—280–345 million years ago—provides evidence that warm and humid conditions prevailed when it was growing. Ferns appeared in the Devonian period and have thrived on Earth for more than 300 million years, retaining the same essential form throughout this time. Their close relatives, the seed ferns, grew in such abundance during the Carboniferous that the earliest coal deposits were formed from their remains.

← **Crater revealed** This crater in the Sahara Desert is one of the ancient meteorite craters uncovered by satellite radar-imaging techniques. The impact of meteorites may have triggered prolonged cool-climate episodes as they threw enough dust into the air to block out the Sun. Crater structures so large that they are visible only from a satellite indicate that giant meteorites have struck Earth several times in its 4600 million-year history. Thousands of smaller meteorites strike the planet each year.

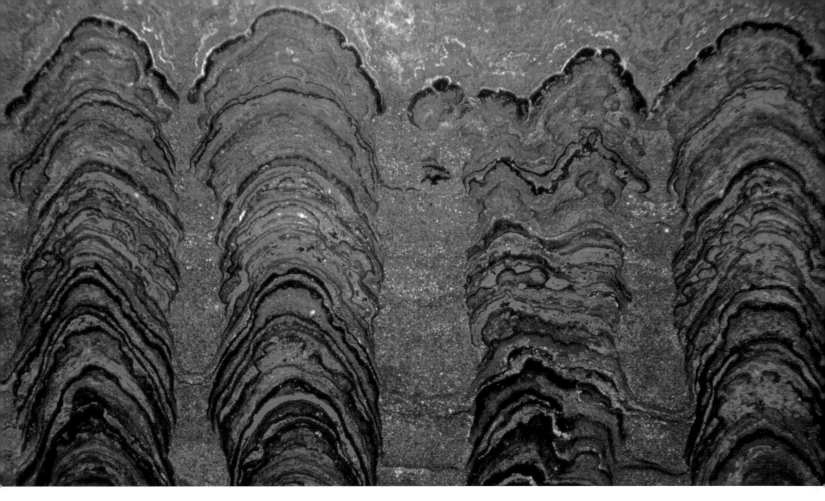

↑ **A great survivor** Stromatolites, cushion-like masses composed of layers of blue–green algae, are among the few organisms that survive from the pre-dinosaur age. By comparing this Argentinean fossil with today's species, estimates can be made of climatic conditions millions of years ago. The oldest stromatolite fossils date back approximately 3500 million years. They lived in the shallows of saline, sunlit waters. Living stromatolites are found in regions as diverse as the hot springs of Yellowstone National Park, USA, the west coast of Australia, and Hudson Bay, Canada.

← **Filter feeder** Analysis of fossils like this crinoid—a member of a family of lily-shaped marine invertebrates—from rocks approximately 300 million years old, indicates that temperatures in the oceans were relatively mild. The crinoid used its feathered arms to filter microscopic forms of life from the sea. It was attached to the sea floor with a stem. The fossil record is substantial but incomplete.

The dinosaur era

About 330 million years ago, at the beginning of the Carboniferous period, temperatures dropped, culminating in a long "big freeze"—the Permo–Carboniferous glaciation—that lasted until 245 million years ago. Then followed a prolonged period of warmth that provided good conditions for plant growth across the mega-continent of Pangea. One theory is that these conditions were brought about by a change in Earth's angle of tilt, combined with the effects of what was to all intents and purposes one large expanse of ocean.

WARM OCEANS
The temperatures of the oceans during the Mesozoic—dinosaur era—were warm, allowing a great variety of marine life to evolve. Warm ocean temperatures are often associated with increased rainfall over continental areas, and vigorous plant growth. This, in turn, helped support the massive, plant-eating dinosaurs.

CHANGING WORLDS
The environment of the Triassic period in which the first dinosaurs lived roughly 228 million years ago would be unrecognizable to us. It was a world in which there was only one major landmass, where there were red, dry landscapes and forests without a single flower. By the end of the Triassic many features of our world were in place. Tectonic forces were tearing the supercontinent apart, plants and animals were evolving, and climates were becoming more seasonal.

← **Ocean fossil**
Plesiosaurs such as this fossilized specimen lived in the seas more than 230 million years ago. The size of these marine predators suggests that the oceans were able to sustain large numbers of creatures over millions of years.

← **Remains tell a tale** These fossilized pterodactyl remains found near Wurtemburg in Germany are similar to those found on other continents. This suggests that favorable climate conditions existed across vast but widely separated areas of Earth during the dinosaur era. Pterodactyls were close relatives of the dinosaurs, and were the first true vertebrate fliers.

→ **When reptiles flew** A lifesize model of a pteradon above Smokey Hills, Kansas, USA, is soaring over chalk deposits that were once the bottom of a shallow ocean, known as the North American Seaway. These deposits have yielded well-preserved dinosaur fossils dating back 100 million years—another clue in understanding Earth's history.

WHAT KILLED THE DINOSAURS?

There has been much discussion about what brought the era of dinosaurs to an abrupt end 65 million years ago. Most agree that it was a sudden, catastrophic change in the climate, but much conjecture remains. Sudden global cooling may have been caused by the impact of a large meteorite or comet. Some believe that other factors were also important, including the drift of large continents into more southerly latitudes. These landmasses cooled dramatically during winter, leading to the formation of polar ice caps, and cooler climates in many parts of the world.

Continental drift would also have changed the circulation pattern of the global ocean currents. Perhaps a land configuration partly blocked the flow of warm equatorial ocean waters into the midlatitudes. This would have produced a significant drop in average temperatures over large areas, perhaps changing the vegetation patterns from the lush, tropical cover that supported the dinosaurs. An event like this would have been devastating, even over a comparatively short period of time.

← **Giant bones** In 1909 Earl Douglas of the Carnegie Museum, Pittsburgh, USA, noticed a dinosaur skeleton eroding out of an exposed sandstone ledge in Utah. Several years of excavation work brought to light a seemingly endless array of dinosaur fossils. Today the site is part of the Dinosaur National Monument. Ten genera have been found at the site. They inhabited an extensive, lowland, alluvial plain and were probably swept away and drowned by large-scale floods. The carcasses would have been dumped at the river bends, where the currents slowed.

Natural climate clues

The most accurate way to measure the climate is to use properly calibrated meteorological instruments that record variables such as temperature, humidity and rainfall. Records obtained using instruments in this way go back about 300 years, which in climatic terms is a very short time. To learn about the climate before that, a certain amount of analytical detective work—relying on the interpretation of a range of naturally occurring events that are weather sensitive—must be used. These phenomena are known as proxy indicators and science is becoming increasingly expert at interpreting them and building up a knowledge of past climates. This is one of the more exciting areas of climatology, and is helping to explain historical events such as past human population movements.

LOOKING INTO THE CLIMATE CRYSTAL BALL

A growing number of proxy indicators are being identified, along with a developing understanding of their significance. Areas of interest to meteorological historians include the study of coral, ice cores, stalactites and stalagmites, tree rings, glaciers, sand dunes, lakes, sea shells and marine sediment. Both corals and trees, for example, have growth rings that provide clues to long-term climate change, while ice sheets contain information about Earth's climate as far back as 420,000 years ago. Science is constantly seeking other sources of stored climate information and this research is slowly providing information about the weather conditions of times past.

→ **The greatest survivor of all** The bristle cone pines that grow at high altitudes on the arid mountains of North America's Great Basin, from Colorado to California, are the oldest known living trees, dating back more than 4000 years. A cross-section of the rings in a gnarled bristle cone trunk provides a valuable insight into the past rainfall of the area.

← **Subterannean rain gauges** Natural climate clues are varied in their location and formation. The stalactites and stalagmites in caves are a useful measure of past rainfall. Wet-weather periods produce fast growth, and dry spells slow growth, so the dating of sections of these sculptural formations can be used to identify such times. Evidence from phenomena such as this is vital in understanding natural changes in Earth's long-term climate.

↑ **The rings that tell a tale** Each ring within the trunk of a tree represents a year of growth, with broad rings indicating favorable years and narrow rings, leaner times. Long-lived, slow-growing trees contain an easy-to-read record of climates long past.

↗ **Coral's changing skeleton** Corals, which can live for many centuries, have a skeleton composed of calcium carbonate. The density of the skeleton varies with climatic conditions, and this has provided some insight into past temperature and rainfall patterns.

→ **The past, trapped in ice** Scientists drill deep into the thick ice that covers the surface of Antarctica to obtain ice cores bearing air bubbles trapped thousands of years ago. When analyzed, these give up their record of past temperature trends. Greenland ice sheets provide data over 15,000 years; ice from Antarctica extends our knowledge for more than 420,000 years.

Evidence from the glaciers

Glaciers are large bodies of ice that move slowly down a slope or valley, or spread outward on a flatter land surface. Their ebbing and flowing provides evidence intrinsically related to Earth's changing climate. In recent times, glaciers have melted so rapidly that some predictions suggest they will have disappeared completely by 2100. As the glaciers continue to retreat, summer water flows will decrease and the water available for irrigation and hydroelectric power will be reduced. Moreover, melting permafrost could increase the frequency of soil erosion and landslides. Evidence of glacial retreat appears to be general, not localized: it appears in the Andes in south America, the Alps in Europe and in the retreating ice cap on Mount Kilimanjaro in Africa. In 1953 Sir Edmund Hillary pitched a base camp on the Himalayas before his ascent of Everest; by 2002 the glacier had retreated by 3 miles (5 km). Glacial melting could directly affect Europe's weather. The Gulf Stream maintains a mild climate in western Europe compared with other countries of the same latitude. If the glaciers and icebergs melted and added a flush of water to this current there could be a dramatic drop in temperature across the North Atlantic.

↓ **The view from above** A glacier on Ellesmere Island, in the high latitudes of the Northwest Territories, Canada. In past ice ages, glaciers covered much of the northern hemisphere. As more of Earth's surface water was bound up in ice sheets, the sea level was lower than today. Land bridges appeared where once there were *seas*.

→ **Ice in perspective** Tourists appear minuscule as they observe the massive Perito Moreno Glacier in the Los Glaciares National Park, Patagonia, Argentina. Glaciers are today found in high latitudes but in past times they were more widespread. At the height of the last ice age, glaciers covered much of North America.

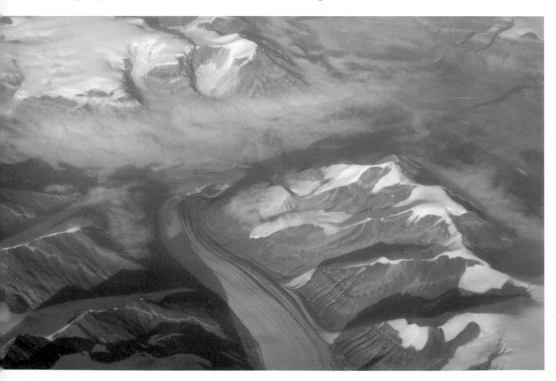

Volcanic skies

Until the 100-day eruption of Krakatoa in 1883, the link between large volcanic eruptions and subsequent adverse weather patterns was not generally acknowledged. The violent eruption of volcanic ash and gases into the upper atmosphere can be responsible for drops in worldwide temperature, lasting, in some cases, for several years. This is because, in a powerful eruption, some of the ash and gas may pass into the stratosphere, where it circulates for several years. Large ash clouds reflect some incoming solar radiation back into space, thus inhibiting the heating effect of the Sun. This conribution to global climate change was first noted by Benjamin Franklin after a volcanic eruption in Iceland in 1783.

THE UNPREDICTABLE INGREDIENT

Massive volcanic eruptions can send vast amounts of ash and sulfur dioxide into the upper atmosphere. The sulfur dioxide reacts with stratospheric water vapor to produce a dense haze that can stay in the stratosphere for years. This haze absorbs some incoming solar radiation and reflects more back out to space, thus raising the temperature of the stratosphere and cooling the lower levels of the troposphere. If the eruptions are large enough, such as the massive Mt Pinatubo eruption in 1991, the effects can last for years.

GLOBAL COOLING

In mid-June 1991 the largest volcanic explosion since that of Krakatoa occurred in the Philippines. Twenty million tons of sulfur dioxide were released by the eruption. Once in the stratosphere, this became sulfuric acid, an aerosol that rapidly spread around the globe, cooling Earth's temperature by an estimated 0.9°F (0.5°C). High levels of aerosols remained in the atmosphere for at least two years, possibly masking the effects of global warming.

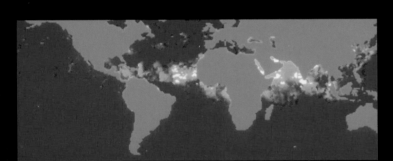

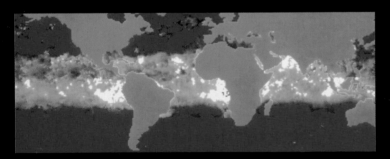

Spreading aerosols Satellite images show the distribution of volcanic aerosols in the atmosphere a week after the explosion of Mt Pinatubo (*above left*)—roughly the normal amount of aerosols, with a slight increase over the Indian Ocean—and then 10 weeks later (*left*). The aerosols then gradually spread to cover the globe, resulting in unusually red sunsets and sunrises. In April 1992, nearly a year after the eruption, the crew of an orbiting space shuttle took the picture below of the Sun rising through a dark band of aerosols from Pinatubo.

A BLAST FROM THE PAST

Fine ash from the eruption of Mt Pinatubo reached to a height of 49,000 feet (15,000 m), penetrating the stratosphere. An estimated two and a half cubic miles (10.5 km³) of ejecta were carried by stratospheric winds around the globe. Volcanic dust also has an impact on our blue skies and red sunsets. The addition of volcanic dust to the atmosphere increases the number of particles and the scattering of colors in the spectrum. Colors at the red end of the spectrum are more effectively dispersed across the sky, producing paler daytime skies, and intense red and purple sunsets.

Location Luzon, Philippines
Latitude and longitude 15.13 N, 120.35 E
Elevation 4900 feet (1485 m)
Volcano type Stratovolcano
Earliest eruption 4100 BC
Oldest historic eruption 1315
Most recent eruption 1992
Number of eruptions in the 20th century 2
Largest eruption 1991

SULFUR DIOXIDE EMISSIONS

It is now thought that sulfur dioxide emissions, rather than ash, have the most dramatic effect on temperature. The eruption at El Chichón, Mexico, which occurred two years after that of Mt St Helens in the United States, released a similar amount of ash, but the sulfur dioxide released was far greater. These emissions were not measured until the 1970s, but it is believed that the emissions from Tambora, south of Borneo, in 1815, must have been immense. A drop in temperature of 3.5–5.5°F (2–3°C) after the eruption of Tambora resulted in drastic food shortages, with accompanying riots in France, famine in Switzerland and crop failures in America. The following year, 1816, was known as "the year without a summer."

RECENT MAJOR VOLCANIC ERUPTIONS

Location	Ejecta (cubic miles [km³])	Sulfur dioxide (million tons [tonnes])
Tambora (1815)	12 (50)	Unknown
Krakatoa (1883)	4 1/3 (18)	Unknown
Mt St Helens (1980)	1/8 (0.5)	1.2 (1.1)
El Chichón (1982)	1/8 (0.5)	7.7 (7)
Mt Pinatubo (1991)	2 1/2 (10.5)	22.4 (20)

Ice ages

Over Earth's long history, ice ages—periods of glaciation—have occurred approximately every 200 million years, and have lasted for millions, or even tens of millions of years. During an ice age, the polar regions are cold, extensive glaciers cover much of the planet, and temperature variations between the poles and the equator are strongly differentiated. We are living in the Cenozoic Era, which has seen three periods of cooling: about 36, 15 and 3 million years ago.

THE HUMAN CONTEXT

The third major period of Cenozoic cooling may have influenced the development of humans. About 2½ million years ago, in sub-Saharan Africa, arid and open grassland expanded, replacing a wetter and more wooded landscape. Paleontologists believe this change in environment is linked to the evolution of humankind. About a million years ago our ancestors spread from Africa to populate the Old World.

↓ **Frozen in time** Antarctica's layers of ice extend our climate record back as far as 420,000 years, beyond the last two ice ages. The thickness of annual ice layers records precipitation, and chemical analysis of the ice reveals the temperature at which the precipitation took place. Air bubbles trapped in the ice record the composition of the atmosphere; dust is a measure of storminess; and acidity reveals major volcanic events. Cores drilled into the ice in Greenland show that about 10,000 years ago the climate was highly variable.

↑ **Sweet survivor** The beautiful liquidambar or sweetgum disappeared from Europe during the ice ages but survived in North America. The rapid cooling of the climate during the Great Ice Age 650,000 years ago wiped out many species of plants and animals on all continents. In Europe, successive waves of ice eliminated species that are found in the United States and China, including not only the sweetgum but also kiwi fruit and the tulip tree, which could not escape south across the east–west barrier of the Alps and Pyrenees.

Buried clues to history The existence of the ice ages was proved by the geological evidence uncovered in the piles of glacial debris—known as moraines—in valleys such as Tracy Arm, Alaska, that were gouged out by vast, slowly moving rivers of ice many thousands of years ago. Glaciers were widespread between 2700 and 1800 million years ago. It then seems that Earth became warmer and free of glaciers for some 800 million years. Beginning about 1000 million years ago, in the late Precambrian Era, there were three distinct periods of glaciation, each of which was a major climatic event. The most recent glacial period reached its peak about 18,000 years ago.

LOUIS AGASSIZ

Louis Agassiz (1807–73), a Swiss naturalist, was the first to propose, in 1837, that "a sudden intense winter, that also was to last for ages, fell upon our globe." He came to this conclusion as a result of a field trip in the Jura Mountains, Switzerland, the year before, when he became convinced that the huge blocks of granite he came across had been transported some 60 miles (100 km) from their point of origin. He reasoned that only glaciers could have done this and that therefore there must have been a much colder epoch affecting the area sometime in the past. Despite initial scepticism from the scientific community, Agassiz's theories have been vindicated.

Recent freezes

The last ice age peaked around 18,000 years ago, when ice sheets up to 10,000 feet (3 km) thick covered most of North America, all of Scandinavia, the Urals and the northern half of the British Isles. In the southern hemisphere, much of New Zealand and Argentina, and the southern part of Australia were iced over. Ice replaced ocean water, sea levels fell, temporary land bridges appeared and humans were able to move into new territory. Conditions became warmer some 7000 years ago, but the process of cooling and warming fluctuates, and relatively minor and short-term changes have occurred since then.

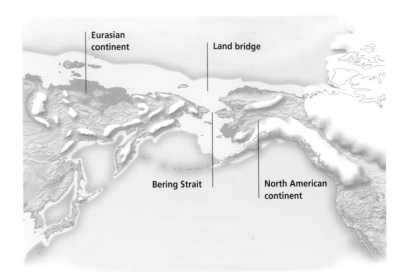

THE LITTLE ICE AGE

From about 1450 to 1850, average temperatures were 2°F (1°C) lower than they are today. Glaciers were extensive in most alpine areas and normally free-flowing rivers froze over in winter. The River Thames in London was the site of frost fairs. A series of cold, wet summers created food crises throughout Europe. Although many regions experienced cooling, the use of the term Little Ice Age has been questioned because it cannot conclusively be considered an event of global significance.

Icy winter Flemish painter Pieter Brueghel (c 1520–69) depicted the harsh winter of 1565 in *Hunters in the Snow*. The so-called Little Ice Age was probably not a single long-term phenomenon, but a series of severe cold intervals.

Walking to another continent The map above shows the land bridge that enabled the settlement of the Americas from Europe. Coastlines and ice sheets are depicted as they were at the peak of the last glacial period, 18,000 years ago.

MOVING WITH THE TIMES

Falling sea levels during periods of glaciation led to the formation of land bridges, which joined what are now neighboring continents or provided a series of "stepping stones" by which people could move from one region to another. Australia's Aboriginal people, for example, migrated from southeast Asia some 50,000 years ago, probably traveling by outrigger canoe, island-hopping along the chain of islands that make up contemporary Indonesia. Thousands of years later, between 25,000 and 14,000 years ago, in the most recent glacial, eastern Siberia was connected to Alaska by a land bridge, allowing humans to cross by land to the New World.

Continents apart
Only some 50 miles (80 km) of icy waters today separate Russia's Chukchi Peninsula from Alaska. This image of the Bering Strait, taken in April 2002, shows the break-up of sea ice between Alaska's Seward Peninsula (on the right) and Siberia (on the left). The Chukchi Sea is to the north.

Climate and civilization I

Human civilization developed some 6000 years ago when a period of warm and stable weather emerged from the cataclysmic fluctuations of prehistoric times, and provided the conditions that fostered agriculture in Mesopotamia and Egypt. Surplus food could be stockpiled, and groups of people gathered together at first in villages which later expanded to become great cities. When conditions were favorable, nations would expand and extend into new lands beyond traditional boundaries. But when less favorable climatic conditions returned, many civilizations declined and often abandoned their newly won territories. These changes were not uniform around the world. Worsening climatic conditions in one part of the world often coincided with improved conditions in another. There is, therefore, a significant link between climate and human migration, a link that has left its imprint on the character and cultures of modern civilization.

↑ **Dreamtime creation figures** The ancestors of Australia's Aborigines migrated from southeast Asia at a time of low sea levels, perhaps 50,000 years ago. Aboriginal rock art provides evidence of changing economic and climatic conditions.

← **In ancient Egypt** Hieroglyphs on the tomb of Horemheb, a king of the eighteenth dynasty of ancient Egypt, pay tribute to the Sun god Ra, in the hope that future River Nile floods, upon which agriculture depended, would be regular. This recognition of the Sun as a life force is common to many cultures. Times of inclement weather, involving too much sunshine (drought) or too little sunshine (flood) indicated an imbalance of natural forces and possibly the deity's displeasure.

WHEN CIVILIZATIONS FLOURISH

Favorable climatic conditions were generally times when rainfall was plentiful and reliable, and temperatures were mild or warm. These conditions are ideal for growing crops and raising domestic animals. Less time was required to find the food and water necessary for survival, allowing more time for exploration and conquest of new territory. An equable climate also meant that communities were better able to support artistic and scientific activities.

The ancient Egyptians depended on the River Nile but did not know its origins or why it flooded regularly. They deified it as Hapi, giving it the form of a well-nourished bearded blue man with female breasts. Because regular flooding of the Nile was so vital, the Egyptians measured its water level in order to predict the harvest. At first these records were little more than marks on the riverbank; later marked stairs, pillars or wells—so-called nilometers—were built. When the Nile failed to rise or when it rose too much, hardship ensued in even the most prosperous periods. And when climatic disturbances in the Nile's catchment area were prolonged, the whole fabric of society might fall apart. Five hundred years of excessive flooding 12,000 years ago caused the Egyptians to abandon their early attempts at agriculture and return to a nomadic existence. Poor floods around 4000 years ago signaled the end of the Old Kingdom; high floods 2000 years later weakened the power of the Middle Kingdom dynasties; and low floods around 1100 BC accompanied the decline of the New Kingdom.

→ **Bathing in the holy Ganges** These Hindu women of northern India carry offerings of food as they bathe in the Ganges River during the Chaith Festival which requires believers to pay homage to the forces of the Sun that influence their daily lives.

Climate and civilization II

Records left by past societies provide evidence of climate change in the human past. Paintings and carvings on cave walls, buildings, written records, archaeological remains—all contribute to our understanding of where and how our forebears lived. Climate played a pivotal role. Cosquer Cave, rich in paintings and stencils created 18,000 years ago, but now submerged under the Mediterranean Sea, is graphic evidence of the flooding that changed the European coastline at the end of the most recent glaciation. Droughts, floods and periods of bitter weather have forced people to abandon their homes and move to more favorable locations.

A COLONY EXTINGUISHED

The collapse of a Norse colony in Greenland provides a case study of the effect of changing climate on a vulnerable human settlement. The colony was established in 985, during a warm period, with 300–400 colonists making the voyage from Iceland. At first, the settlement prospered and by the early twelfth century it supported about 5000 people. Supplies were brought in from Iceland. But then the weather cooled, storms intensified and the pack ice expanded. Visits from Iceland became less frequent. The last contact was in 1410 and the settlement died out later that century.

↑ **The long winter** These ruins are all that remain of the Icelandic colony in Greenland that was wiped out by rapid climate cooling in the early fifteenth century. Archaeological evidence presents a harrowing picture of malnutrition and suffering as the colony struggled against increasingly bitter weather.

← **A record of better times** These frescoes at Tassili Najjer in Algeria, estimated to be between 3500 and 6000 years old, frequently depict the herding of cattle. This suggests that local conditions in this now-arid area were once fertile and productive. Evidence such as this builds up our understanding of historical climate change.

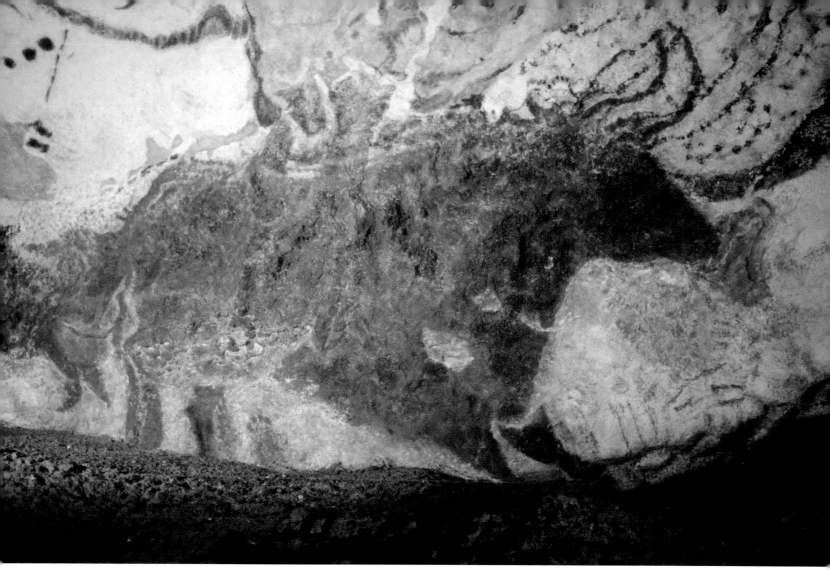

↑ **Ice age art** Cave art appeared in Europe about 24,000 years ago and was at its peak at the height of the last glaciation, around 18,000 years ago. The warmer climate near the Atlantic coast in southwestern Europe enabled people to be less nomadic, as fish and plant foods were abundant. Animals long extinct populated their world: woolly mammoths, wild horses, cave lions and steppe bison. This representation of a bison is a reconstruction from a frieze in Lascaux II, Dordogne region, France.

← **The vanishing Mayans** The Mayan civilization in Central America declined between 800 and 1000 AD. One reason for this may have been crop failure caused by extended drought that resulted from warmer weather. Once-thriving centers such as Chichen Itza, in Mexico, were abandoned.

The enhanced greenhouse effect

So-called greenhouse gases—among them water vapor, ozone and carbon dioxide—make Earth habitable by absorbing infrared radiation and re-radiating it to Earth, thus warming the planet and providing the conditions that enable life to exist on Earth. In postindustrial times, however, these natural greenhouse gases have become more concentrated, with a measurable increase in the amount of carbon dioxide, methane and nitrous oxide in the atmosphere. The natural greenhouse effect has therefore become enhanced, and with it Earth has become warmer.

In 2001, the United Nations-based Intergovernmental Panel on Climate Change (IPCC) reported that demonstrable changes in the global climate had occurred "since the pre-industrial era, with some of these changes attributable to human activities." Widespread concern has led to considerable debate but a fully coordinated global response to such phenomena as the enhanced greenhouse effect has yet to evolve.

THE HUMAN ELEMENT

The enhanced greenhouse effect refers to the contribution to global warming that may have been caused by human activity. Greenhouse gases that have been vented into the atmosphere by various human activities include carbon dioxide, methane, nitrous oxide, hydrofluorocarbons, perfluorocarbons and sulfur heaxafluoride. Atmospheric concentrations of carbon dioxide have been steadily increasing since the start of the Industrial Revolution in the 1800s, which produced the mass burning of fossil fuels to power industrial processes. Scientists estimate the buildup of greenhouse gases by developing global climate models that simulate conditions for the equivalent of many years. If carbon dioxide emissions are not curtailed, such models suggest that they will double by 2060. Based on climate modeling, scientists predict that this would lead to an increase of between 3° and 8°F (1.5°–4.5° C) in average global temperatures, with polar regions warming by as much as 16°F (9°C). The effects are yet to be determined.

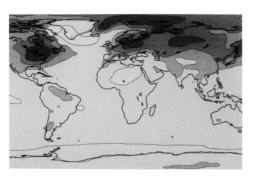

ATMOSPHERIC CARBON DIOXIDE

Keeling Curve from Mauna Loa, Hawaii

← **Measuring carbon dioxide** This graph shows the carbon dioxide measurement from the atmospheric monitoring station at Mauna Loa, Hawaii, USA, from 1920 to 2000. Concentrations of the gas have been increasing at an accelerating rate since the 1800s; from a pre-industrial level of around 280 parts per million (ppm) they had reached almost 370 ppm by 2002. The 30 percent increase is attributable almost entirely to human activities.

→ **Nature's balance** Trees play a vital role in maintaining a healthy atmosphere. They absorb carbon dioxide and store it in their tissues as carbon compounds. When forests are cleared, the carbon is released back into the atmosphere as carbon dioxide. Moreover, land clearing alters the reflectivity, or albedo, of Earth's surface. Long-term climate effects remain undetermined.

← **Modeling the climate** This climate model predicts the effect of doubling current levels of carbon dioxide in the atmosphere. Temperature increases are from pale yellow to red. Dramatic warming occurs across all continents and the Arctic Circle in the northern hemisphere.

The human impact I

Forests are Earth's "lungs," converting carbon dioxide into life-supporting oxygen during the process of photosynthesis. Over the last century or so, these lungs have become increasingly threatened, partly as a result of rapid population growth. Forests have been razed to provide timber and farmland. Conversely, the spread of towns and cities has consumed valuable agricultural land and increased the levels of atmospheric pollution. Increasing numbers of livestock in semi-arid lands have accelerated the process of desertification. All these factors combine to alter local, regional and global climates. To put it simply, as our population continues to increase, climatic changes will become more marked and the delicate balance of plant and animal habitats will be increasingly disrupted.

CAUSES OF THE CREEPING DESERT

Contrary to popular belief, droughts are not the primary cause of desertification. Droughts are quite common in arid lands, and ecosystems have evolved that recover quickly once sporadic rains arrive. The main cause of the spread of deserts is long-term, poor management of marginal lands. Nowhere has the desertification process been more pronounced than in the Sahel region of west Africa. The combination of bad land management and a drought that began in 1968 led to the deaths of more than 100,000 people and 12 million cattle. The Sahel is the transition zone between the Sahara to the north and the tropical forests of equatorial Africa. Its desertification has created a southward shift of the Sahara.

→ **A mixed blessing** An outstanding example of how the world's burgeoning need for food is increasing the pressure on global climate is the clearing of heavily forested mountain slopes for the planting of methane-producing rice fields.

↓ **Disappearing forest** Remaining tropical forest in Brazil is bright red, while dark red and brown areas represent cleared land, and black and gray patches are recently burned. The vertical lines indicate land cleared along transport routes.

Driving into the wilderness Many scientists are seriously concerned about the building of roads through the Amazonian forests of Brazil. Improved road access leads to increased clearing of the remaining rain forests and, in turn, more roads.

THE UNSTOPPABLE HUMAN RACE

Never before has Earth been home to so many people. It took most of human history to reach the billion mark, around 1800. Since then, growth has accelerated rapidly, with the sixth billion, achieved in 1999, added in a mere 12 years. United Nations population projections point to the population increasing to around nine billion by 2050. The corresponding demands for food, water, shelter and transportation will inevitably increase the pressure on Earth's resources and climate. Sustaining ever-increasing populations while combating environmental issues such as desertification, deforestation and urban sprawl has become a challenge of some magnitude, particularly as the most rapid growth is in developing countries.

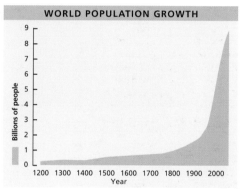

WORLD POPULATION GROWTH

When the desert takes over The culture of the cattle-rearing Wodaabe people of the west African state of Niger is under intense and growing pressure as a result of desertification of their semi-arid environment.

Precious commodity Awareness that the water supply for agriculture is finite has led to changes in the traditional flood-irrigation practices used for growing cotton. Maintaining the availability of good-quality water is now a global priority.

The human impact II

For most of Earth's history, human activity was on such a small scale that its effect on the planet's climate was minimal. But in more recent times, massive advances in agriculture, industry and transport have had unwanted consequences. When forests are cleared, the trees that absorb carbon dioxide are lost. Automobile exhausts and the burning of fossil fuels release carbon dioxide into the air.

THE ONGOING INDUSTRIAL REVOLUTION

Industrial processes release large amounts of carbon dioxide and other gases into the atmosphere and some of these react with water vapor to produce acid rain, which is known to kill off large tracts of forest. Air-quality monitoring stations have measured increasing levels of so-called greenhouse gases, the origins of which can be traced to industry. As emissions increase, so man-made gases continue to accumulate in the atmosphere.

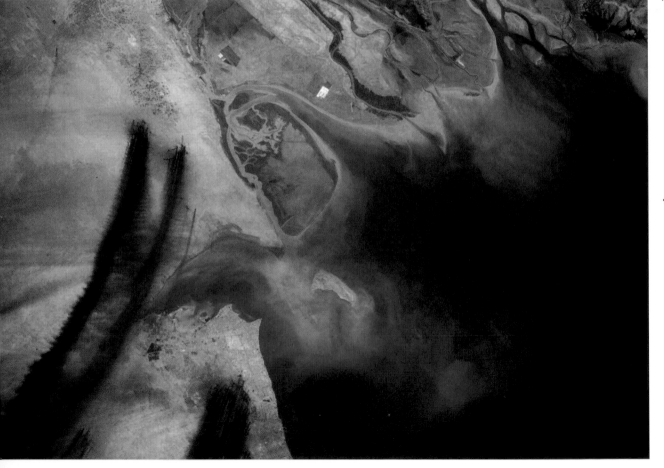

↑ **Friend or foe?** This complex Tokyo freeway system is a potent symbol of one of the major sources of carbon-dioxide greenhouse gases—the automobile. The increasing use of motor vehicles globally poses a serious threat to future climatic conditions.

← **A different kind of casualty** These oil fires across Iraq and Kuwait during the Gulf War of 1991 released massive amounts of toxic pollutants into the atmosphere. The yellow color is the desert sands; the blue the Persian Gulf waters. Thick black smoke from more than 700 oil well fires is being carried from Iraq toward Saudi Arabia, leaving a pollution trail across the desert. Pollution of this kind is a casualty of modern desert warfare.

→ **Gas producers** These Bos Taurus cattle grazing the Brazilian savanna expel methane gas that is a major contributor to the gases that cause the greenhouse effect.

→→ **Mixed blessing** Exhaust gases from high-flying aircraft are able to reach the stratosphere very easily, altering the chemical balance in this layer.

TRYING TO BE CLEANER AND GREENER

Because the burning of fossil fuels to produce energy releases carbon dioxide—a greenhouse gas—the use of so-called green power is being encouraged by some governments and is gradually being seen as more attractive by many industries. Environmentally friendly power developments include solar energy, wind turbines and even devices that harness the energy of ocean waves and tides. Global will is needed to maximise the potential of these alternatives.

Pick the culprit The cooling-tower emissions of this West Virginia, USA, power station are mostly harmless water vapor. Polluting carbon dioxide streams from the narrow chimneys.

Ozone depletion

Ozone—a gas that in the stratosphere absorbs ultraviolet (UV) rays from the Sun—plays a vital role in protecting life on Earth from the harmful effects of UV radiation. High levels of exposure to UV light cause skin cancers in humans and also harm many other forms of life. Since the late 1970s, scientists have noticed that each year during late spring a large area of severely depleted ozone, known as the "ozone hole," forms over the Antarctic continent; a smaller hole appeared for the first time over northern polar skies in the late 1990s. The size of these holes is increasing each year. Investigations have shown that man-made chemicals—chlorofluorocarbons (CFCs)—rise high into the stratosphere and deplete the ozone layer. An international treaty— the Montreal Protocol—was signed in 1987 (and later amended) to eliminate certain CFCs from industrial production. Subsequently, global use of the most harmful CFCs fell by 40 percent within five years but it will be decades before the CFCs in the atmosphere are eliminated.

→ **Twin eruptions** Contrasting ash clouds were spewed into the atmosphere when two volcanoes erupted spontaneously at Rabaul in Papua-New Guinea in 1994. Dust and ash from volcanic eruptions contribute to ozone decline.

↘ **Sunspots: complicating the issue** Natural stratospheric ozone concentrations follow the 11-year oscillation of sunspot activity, making annual predictions of the size of the ozone hole more difficult. These sunspots were observed in July 2002.

↓ **Protection from the Sun** A growing public awareness of the harmful effects of unprotected exposure to the Sun, and the increased potential for skin cancer with the decline in the ozone level, has led to more appropriate beachwear.

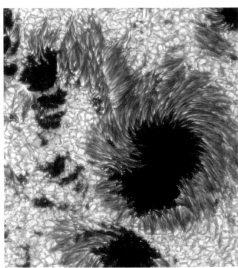

HOW THE OZONE LAYER IS PUNCTURED

CFCs interfere with the natural process of ozone formation when they reach the stratosphere and, in a process of chemical breakdown and change, form chlorine compounds that destroy ozone. This occurs over Antarctica at the end of each winter, when an intensely cold vortex creates ice clouds and traps CFCs. When the Sun returns, the combination of sunlight, iceclouds and CFCs forms an ozone-destroying mixture.

↑ **Antarctic research station** The Dumont d'Urville research station is one of several in Antarctica undertaking research into ozone depletion. Folklore has it that when the first measurements were taken in 1985, the drop in ozone levels was so dramatic that the scientists thought their instruments were faulty. Replacement instruments confirmed the first measurements several months later, and the ozone depletion was accepted as genuine.

A WORRYING "HOLE" IN THE SKY

The ozone hole over Antarctica has expanded considerably over the last few decades. In the illustration the dark shades of blue show low levels of stratospheric ozone. It is only over the last couple of years that a slowing of the expansion of this hole has become evident. This slowdown has been hailed by some scientists as a sign that the reduction of CFC production that followed the international treaty of 1987 is beginning to bear fruit. Most ozone is produced over the tropics, where solar radiation is strongest, and is transported around the Earth by high-level winds. In winter and spring, an isolated air mass forms over the Antarctic, with air flowing around the South Pole. During the long winter, the air is not heated by the Sun and clouds of ice and frozen particles form in the stratosphere. If CFCs are present, these particles interact when spring returns to cause a series of complex chemical reactions that deplete the ozone concentration.

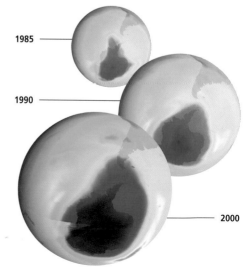

1985

1990

2000

Global warming

Most scientists now agree that global warming must be regarded as a fact rather than the theory it was only a few decades ago: the global climate has warmed by 1°F (0.5°C) in the past hundred years. Accurate temperature measurements and analysis of historical records all show a marked global warming trend since the late 1800s: the 10 warmest years of the twentieth century occurred between 1985 and 2000. The warmest global year yet recorded was 1998.

This evidence is also supported by a growing number of indirect observations. Glaciers are in retreat, shrinking more than at any time in recorded history. Snow lines are creeping higher up the slopes of mountains. Rising sea levels are also consistent with increasing temperatures: as ice melts, oceans expand. Changes in vegetation are also becoming evident, particularly in polar and semi-arid regions.

THE GLOBAL WARMING DEBATE

There is no disputing the fact that Earth is heating up. How much is natural and how much is caused by humans is where the debate begins. Paleoclimatological records show there have been large variations in climate throughout Earth's long history. State-of-the-art global climate models have linked recent warming with increased levels of carbon dioxide in the atmosphere.

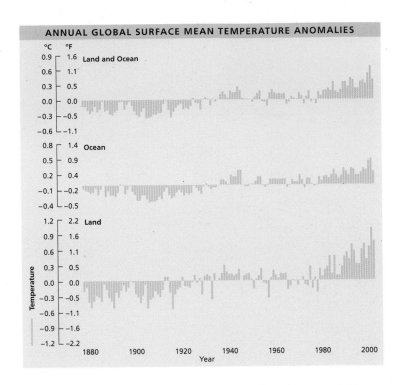

ANNUAL GLOBAL SURFACE MEAN TEMPERATURE ANOMALIES

↓ **Retreat of the tropical snows** Few places show the effects of retreating snow lines more clearly than the slopes of Mount Kilimanjaro, Kenya. The contrast between the hot plains and the snow-clad mountaintop becomes less dramatic each year: the snow cover has shrunk by more than 80 percent since it was first measured in 1912. Climate models project that snow extent and sea-ice will continue to decrease, and that the retreat of glaciers and ice caps will continue.

↑ **The statistics: our planet is warming up** Scientists have drawn up reliable graphs of Earth's temperature trends for the last 120 years, covering the entire globe (*top*), the oceans (*center*) and the land areas (*bottom*). The warming trend in all three temperature plots is evident, with the largest increase occurring over the land. While there have been many periods of so-called abnormal weather, the overall patterns are readily discernible.

WHEN POLAR ICE TURNS TO WATER

Much of the world's water is contained in the polar ice caps and the ice shelves over the adjacent oceans. Although there are seasonal changes around the edges of the ice shelves, the break-up of permanent shelves suggests that something is amiss. As large sections of the ice shelves melt, global sea levels rise.

↑ **The shrinking glaciers** Satellite photographs of these Himalayan glacial lakes—blue ribbons below white glaciers— reveal their growth in recent years, providing tangible evidence of global warming. The melting of mountain glaciers has proceeded at such a rate that some predictions suggest that the majority of glaciers will have disappeared by 2100.

→ **Fraying around the edges** This vast crack in Antarctica's Larsen B ice shelf opened up in February 1997. A 27-mile (43 km) by 3 mile (4.8 km) chunk of the shelf later broke away.

El Niño | La Niña

There is growing evidence that changes in the surface temperature of the tropical Pacific Ocean control many features of global weather patterns, including floods and droughts far from the tropical Pacific. Unusually warm currents—known as the El Niño—are associated with altered atmospheric circulations that extend over large parts of the globe. These include a breakdown of the usual southeast trade winds, which are replaced with a regime of westerly wind. In turn, these tend to drive warm surface ocean waters across to the eastern side of the tropical Pacific Ocean. This triggers floods across the northwest of South America and droughts over much of Australia, and, in extreme cases, Indonesia. Significant but smaller effects are felt in other parts of the world, such as anomalies in winter temperatures in North America and winter rainfall in northwestern Europe. El Niño events occur every two to seven years and can last between three or four years. When the opposite occurs and the currents become unusually cold—the phenomenon called La Niña—dry conditions return to Peruvian coastal regions and floods become likely across eastern Australia.

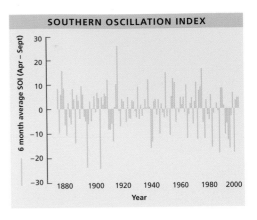

HOW AN OCEAN CURRENT WAS NAMED

The sporadic warm ocean current often arrives off the coast of Peru close to or during the Christmas period. Because of this timing, the local Peruvian fishermen call this current El Niño (Spanish for "the child"), a direct reference to the Christ child. When a colder-than-normal current extends northward past Peru, it is dubbed with the feminine usage, La Niña. As we can predict El Niño events several months ahead, the potential of basing long-term forecasts on El Niño is being explored.

Ocean temperatures   Ocean temperature anomalies

El Niño January–March 1998

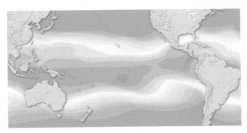
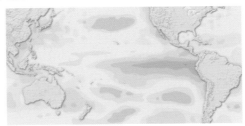

La Niña January–March 1989

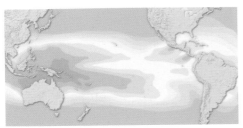

64°F   75°F   86°F
18°C   24°C   30°C

−5.4°F   0°F   +5.4°F
−3°C   0°C   +3°C

WHAT IS THE OSCILLATION INDEX?

The mean sea-level pressure difference between Tahiti and Darwin, Australia, is known as the Southern Oscillation Index (SOI). When the SOI is strongly negative for several months, an El Niño is said to occur; when it is strongly positive, a La Niña occurs.

THE OCEAN'S TEMPERATURE IS THE KEY

The best way to identify whether an El Niño or a La Niña is developing is to monitor the sea-surface temperature patterns across the tropical Pacific Ocean. During strong El Niños, sea-surface temperatures become unusually warm over the equatorial eastern Pacific Ocean, particularly around coastal Peru. This movement of warm water along the coastline is dreaded by local fishermen because the water lacks nutrients and results in decimation of fish across the area, including one of the main harvests—the anchovies. Sea birds that feed off the anchovies also die during these periods. A strong El Niño of this type occurred in 1998 and can be identified in the Ocean Temperature Anomalies diagram (*top left*) in which "anomalies" refer to the variation from average conditions.

The opposite effect occurs during a La Niña, with colder than normal waters making a return to the tropical eastern Pacific and warmer waters being carried on the current north of Australia. A significant La Niña event occurred in 1989 (*bottom left*). La Niña produces the opposite climate variations from El Niño. For example, the parts of Australia and Indonesia that are prone to drought during El Niño are typically wetter than normal during La Niña.

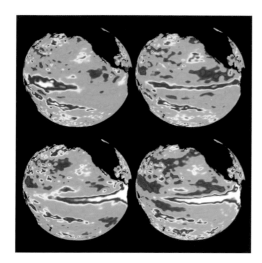

EL NIÑO ON THE MOVE

These time-lapse satellite images show El Niño developing in the Pacific Ocean between April and September 1997. The difference in average sea heights and those caused by warm water of the El Niño are color coded, from purple (4 inches [10 cm] below normal level) through blue, green, yellow and red to white (8 inches [20 cm] above normal). Landmasses are black and the El Niño is the red and white tongue moving eastward across the Pacific along the equator. El Niño events are not caused by global warming; a variety of evidence (including archaeological) indicates that the phenomenon has existed for hundreds of years—and some would argue, millions. However, warmer global sea surface temperatures may enhance El Niño, and they have been more frequent and intense in recent times. Current climate models suggest that El-Niño like sea surface temperature patterns in the tropical Pacific are likely to become more persistent.

↗ **When wildfires strike** The severe droughts associated with El Niño produce the hot, dry conditions ideal for the spontaneous outbreak of often devastating wildfires.

→ **El Niño's bitter legacy** A farmhouse in Ecuador is stranded amid floodwater caused by an El Niño event in 2002. As well as the human toll, floods, wildfires and droughts caused by El Niño have cost millions of dollars' worth in damage to the agricultural and pastoral economies of regions affected by these tropical currents.

Earth facts

EQUINOXES

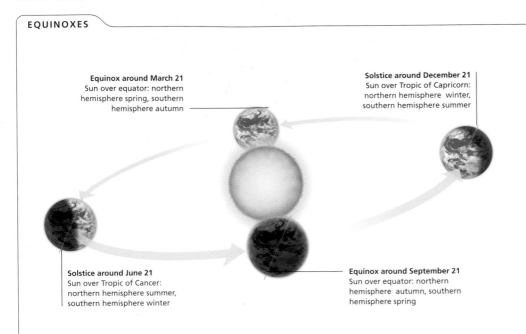

Equinox around March 21
Sun over equator: northern hemisphere spring, southern hemisphere autumn

Solstice around December 21
Sun over Tropic of Capricorn: northern hemisphere winter, southern hemisphere summer

Solstice around June 21
Sun over Tropic of Cancer: northern hemisphere summer, southern hemisphere winter

Equinox around September 21
Sun over equator: northern hemisphere autumn, southern hemisphere spring

EARTH'S COMPONENTS

Atmosphere A gaseous envelope surrounds Earth, protecting life from the harshness of space. The atmosphere's clouds, suspended particles and gas extend at least 60 miles (100 km) above the surface.

Hydrosphere The liquid water component of Earth's system consists of the oceans and other large bodies. The hydrosphere covers about 71 percent of the surface and contains most of the planet's water.

Cryosphere The component of the planet made up of ice is known as the cryosphere and includes glaciers and polar ice caps. A large proportion of the planet's freshwater is found in the cryosphere.

Lithosphere The lithosphere is the solid portion of Earth, and includes the soil and rocks upon which we live. Nutrients from the atmosphere are fixed in the soil and used by plants in the biosphere.

Biosphere As far as is known, Earth is unique among planets because its system supports life. The biosphere is made up of animals, plants and other organisms.

Biosphere interactions Green plants contain chlorophyll, which helps to convert solar energy into carbohydrates through photosynthesis. Oxygen is released into the atmosphere as a byproduct.

THE WATER CYCLE

Evaporation of water to form cloud

Rain falls from cloud

Rivers drain into the ocean

Inland water storages are filled

Subterranean water returns to ocean

THE WATER CYCLE

The oceans cover 71 percent of the globe and contain 97 percent of Earth's water. This seawater is salty and undrinkable. The other 3 percent is fresh water. More than 2 percent of Earth's total water is contained in freshwater ice sheets and glaciers, and less than 1 percent consists of groundwater. A small fraction of a percent is contained in lakes, rivers and the atmosphere. Water is continually being transferred between the oceans, land, plants and the atmosphere in a process called the water, or hydrologic, cycle.

Most water that is in the the atmosphere has evaporated from the oceans. When moist air rises and cools, clouds form. Clouds can produce precipitation if they grow sufficiently. Precipitation either soaks into the ground; runs off into lakes, streams or rivers; or evaporates into the air.

Plants draw water up from the soil via their roots. Eventually, the water passes through the leaves into the air, a process known as transpiration. Water also evaporates from lakes and rivers, falls back to Earth as precipitation, and is absorbed by the ground. The water cycle is completed when water returns to the oceans via rivers and underground streams.

Wind facts

THE BEAUFORT SCALE

	Wind speed (mph [km/h])	Description	Effects on land
0	below 1 (below 2)	calm	smoke rises vertically
1	2–3 (3–5)	light air	smoke drifts slowly
2	4–7 (6–11)	light breeze	leaves rustle; vanes begin to move
3	8–12 (12–19)	gentle breeze	leaves and twigs move
4	13–18 (20–29)	moderate breeze	small branches move; dust blown about
5	19–24 (30–38)	fresh breeze	small trees sway
6	25–31 (39–51)	strong breeze	large branches sway; utility wires whistle
7	32–38 (52–61)	near gale	trees sway; difficult to walk against wind
8	39–46 (62–74)	gale	twigs snap off trees
9	47–54 (75–86)	strong gale	branches break; minor structural damage
10	55–63 (87–101)	whole gale	trees uprooted; significant damage
11	64–74 (102–120)	storm	widespread damage
12	above 74 (120)	hurricane	widespread destruction

The Beaufort scale was devised to measure wind force on sailing ships.

THE SAFFIR–SIMPSON SCALE

	Pressure (hectopascals)	Wind speed (mph [km/h])	Storm surge (ft [m])	Damage
1	more than 980	74–95 (118–152)	4–5 (1.2–1.6)	minimal
2	965–980	96–110 (153–176)	6–8 (1.7–2.5)	moderate
3	945–964	111–130 (177–208)	9–12 (2.6–3.7)	extensive
4	920–944	131–155 (209–248)	13–18 (3.8–5.4)	extreme
5	less than 920	more than 155 (248)	more than 18 (5.4)	catastrophic

Since the 1970s, the National Hurricane Center in the United States has used the Saffir–Simpson scale to classify hurricanes. They are graded in intensity from 1 to 5.

THE FUJITA SCALE

Scale	Speeds (mph [km/h])	Damage
F0	40–73 (64–117)	light
F1	74–112 (118–180)	moderate
F2	113–157 (181–251)	considerable
F3	158–206 (252–330)	terrible
F4	207–260 (331–417)	severe
F5	more than 260 (417)	devastating

The Fujita scale provides a measure of the strength of a tornado.

Ocean facts

OCEAN CURRENTS

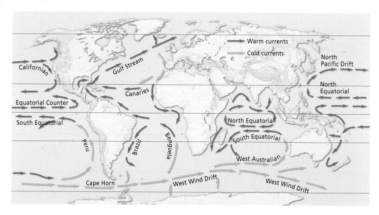

Ocean currents are largely caused by the prevailing winds. Winds from the high-pressure systems in the midlatitudes drive currents that circulate in a clockwise direction in the northern hemisphere and counterclockwise in the south. The highs produce cold currents on the western sides of continents and warm currents on the eastern coasts. The main cold ocean currents, such as the Peru current and the Benguela current, tend to reduce rainfall over nearby land: the cool air above the current holds little moisture, so rain clouds are unlikely to form. The main warm currents, such as the Gulf Stream and the Californian current, have the opposite effect: they produce warmer temperatures over nearby land. Large changes in sea-surface temperatures to the west of these currents favor the development of low-pressure centers.

THE GREAT OCEAN CONVEYOR BELT

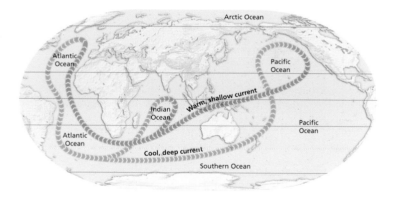

The "great ocean conveyor belt" may be the key to long-term changes in the surface temperatures of the oceans, which are an important factor in global climate. Cold, salty water, which sinks into the deep ocean in the north Atlantic, flows south and then east around southern Asia to resurface and be warmed in the Indian and north Pacific oceans. Surface currents carry warmer water back through the Pacific and south Atlantic. The round trip takes between 500 and 2000 years. The latest studies suggest that the strength of this transport can easily change speed or direction. Recent changes in ocean-water temperature may have contributed to climatic fluctuations, such as the sustained drought in the Sahel since the late 1960s, reduced hurricane activity in the Atlantic and a rise in El Niño events in the tropical Pacific.

Atmosphere facts

CLOUD CATEGORIES

English scientist Luke Howard (1772–1864) was the first to classify clouds, using Latin names to describe three main cloud shapes: cumulus, stratus and cirrus. As well as grouping clouds by their shape, meteorologists also classify them according to their altitude. The first division of the atmosphere into three levels was proposed in 1803 by French scientist Jean-Baptiste Lamarck (1744–1829). Today, clouds are described using a combination of Howard's and Lamarck's systems.

BASIC TYPES

Alto Derived from the word high, but in meteorology used to refer to middle-level clouds.

Cirrus Meaning filament of hair and used to identify high-level clouds.

Cumulus From a term meaning pile or heap, here used to refer to a "tall" cloud of great height.

Nimbus Meaning rain, so used to refer to rain-bearing clouds; commonly used as a suffix.

Stratus Derived from stratum or layer, stratus refers to low-level clouds; also used as a suffix to a set of cloud types that have a layered appearance.

SPECIFIC TYPES

Cirrostratus A combination of cirrus and stratus. Cirrostratus are generally recognizable by a transparent thin white sheet or veil of ice crystals forming high-level clouds that appear as layered streamers.

Cirrocumulus A combination of cirrus and cumulus. High-level ice crystal clouds consisting of a layer of small white puffs or ripples.

Altostratus Stratiform clouds with the "alto" prefix to indicate middle-level altitude. Altostratus clouds consist primarily of water droplets that appear as a relatively uniform white or gray layered sheet.

Altocumulus A middle-level cloud type that has some vertical development as indicated by the suffix "cumulus." Altocumulus clouds generally have a layered appearance but they also consist of white to gray puffs or waves.

Stratocumulus Low-level layer clouds as suggested by "strato" but having some vertical development as indicated by the suffix "cumulus." Stratocumulus clouds consist of a layer of large rolls or merged puffs.

Cumulonimbus Vertically developed clouds (indicated by the "cumulo" prefix) that are also rain producers (indicated by the "nimbus" suffix). These "tall," high clouds usually extend up to the troposphere and have a puffy lower portion and a characteristic smooth or flattened anvil-shaped top. These clouds usually produce heavy rain or hail.

Nimbostratus Rain-producing (nimbus) layered (stratus) clouds. Nimbostratus are low- to mid-level clouds that have the appearance of a uniform gray layer.

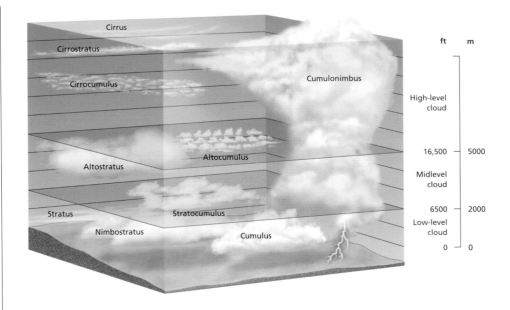

THE AIR WE BREATHE

The air we breathe is in the troposphere, the lowest level of the atmosphere. Nitrogen and oxygen are its major components, with a small amount of water vapor, argon, carbon dioxide and other gases. The water vapor in the troposphere is responsible for much of the weather we experience.

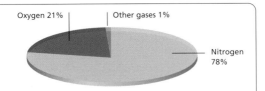

TYPES OF PRECIPITATION

Rain represents liquid precipitation that falls to the ground from clouds. Most rain falls from nimbostratus and cumulonimbus clouds. These rain-producing clouds are sufficiently deep to develop large raindrops. Much smaller drops may fall as drizzle from some stratus or cumulus clouds that are not deep enough to produce heavy drops.

Drizzle usually appears as a fine mist that falls or drifts so slowly through the air that it can appear almost suspended in the atmosphere. Even if drizzle persists for an extended period, only small rainfall totals are recorded on the ground. Rain, on the other hand, can be light or extremely heavy, producing widespread floods and disruption to everyday life in urban areas.

Snow falls when a sufficiently thick layer of air near Earth's surface is at or below the freezing point of water. During winter, snow in the form of snowflakes can fall from stratus- or cumulus-type clouds. No two snowflakes are the same.

Hail is a product of thunderstorms, where ice particles form when supercooled water droplets pass repeatedly through cloud layers of different temperatures. It is perhaps the most destructive form of precipitation and varies in size from pea-sized to orange-sized missiles.

Ice pellets are produced when rain falling through a warm layer of air encounters a deep layer of subfreezing air that turns raindrops solid. Occasionally, ice pellets may accumulate on the ground.

Freezing rain consists of supercooled water droplets that freeze on or near the ground. It can produce ice pellets or a smooth coating of ice known as glaze.

Sleet is a mixture of rain and snow.

ATMOSPHERE LAYERS

The layers of the atmosphere are defined according to their temperature profile. Their heights are approximate and vary considerably with latitude and season. In order to show more detail, the proportions of the layers have been distorted in the illustration below.

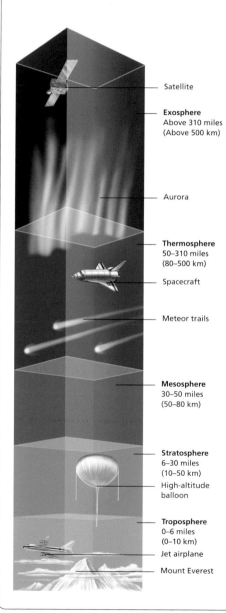

Satellite

Exosphere
Above 310 miles
(Above 500 km)

Aurora

Thermosphere
50–310 miles
(80–500 km)

Spacecraft

Meteor trails

Mesosphere
30–50 miles
(50–80 km)

Stratosphere
6–30 miles
(10–50 km)

High-altitude
balloon

Troposphere
0–6 miles
(0–10 km)

Jet airplane

Mount Everest

INTERNATIONAL WEATHER SYMBOLS

CURRENT WEATHER

light drizzle	steady, moderate rain	hail
steady, light drizzle	intermittent, heavy rain	freezing rain
intermittent, moderate drizzle	steady, heavy rain	smoke
steady, moderate drizzle	light snow	tornado
intermittent, heavy drizzle	steady, light snow	dust storms
steady, heavy drizzle	intermittent, moderate snow	fog
light rain	steady, moderate snow	thunderstorm
steady, light rain	intermittent, heavy snow	lightning
intermittent, moderate rain	steady, heavy snow	hurricane

LOW CLOUDS

stratus	cumulus	cumulonimbus calvus
stratocumulus	cumulus congestus	cumulonimbus with anvil

MIDDLE CLOUDS

alstostratus	altocumulus	altocumulus castellanus

HIGH CLOUDS

cirrus	cirrostratus	cirrocumulus

SKY COVERAGE

no clouds	four-tenths covered	seven- to eight-tenths covered
one-tenth covered	half covered	nine-tenths covered
two- to three-tenths covered	six-tenths covered	completely overcast

WIND SPEED mph (km/h)

calm	9–14 (14–23)	55–60 (89–97)
1–2 (1–3)	15–20 (24–33)	119–123 (192–198)
3–8 (4–13)	21–25 (34–40)	

Temperature and humidity facts

TEMPERATURE

CONVERTING TEMPERATURES

From Celsius to Fahrenheit: °F = (1.8 x °C) + 32
From Fahrenheit to Celsius: °C = 0.56 x (°F – 32)

WIND CHILL

Wind-chill tables show the apparent temperature produced by the combination of actual temperature and wind speed.

	Actual temperature °F				
Wind speed mph	**40**	**30**	**20**	**10**	**0**
15	23	9	– 5	–18	–31
20	19	4	–10	–24	–39
25	16	1	–15	–29	–44
30	12	–2	–18	–33	–49
	4	**-1**	**-7**	**-12**	**-18**
24	5	13	21	28	35
32	-7	-16	-23	-31	-39
40	-9	-17	-26	-34	-42
48	-11	-19	-28	-36	-45

MEASURING HUMIDITY

Air is saturated—containing a high level of water vapor—when a dynamic equilibrium is attained between the rates of evaporation and condensation. The air temperature at which saturation occurs is called the dewpoint. In this graph, if an air mass holds ½ cubic inch of water vapor per cubic yard of air (10.7 cm³/m³), the dewpoint will be 52.5°F (11.4°C).

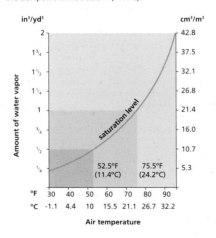

MEASURING THE WEATHER

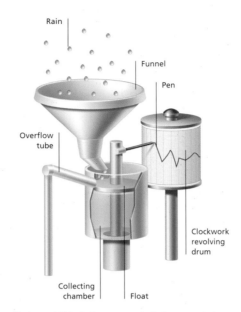

Pluviograph This device measures both the amount of rainfall and its rate. Rain falls into the funnel, runs into a collecting chamber and causes a float to rise.

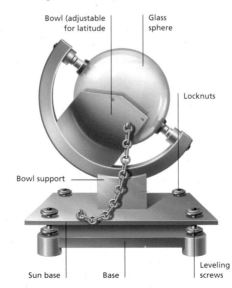

Sunshine recorder A glass sphere concentrates the Sun's rays onto a card to produce a scorch mark.

WEATHER INSTRUMENTS

Wet-bulb thermometer measures temperature of the surrounding saturated air
Dry-bulb thermometer measures actual air temperature
Barometer measures air pressure
Anemometer measures wind speed and direction
Rain gauge measures precipitation
Thermograph records temperature graphically
Hygrograph records relative humidity graphically
Hygrometer measures relative humidity
Pluviograph records rainfall graphically
Wind vane indicates wind direction
Wind sock indicates wind direction and provides a visual clue to wind speed
Sunshine recorder measures hours of sunlight in day and extent of cloud cover in the period
Radiosonde balloon measures temperature, air pressure and humidity vertically in the atmosphere
Ground-based radar measures the position, amount and density of rain and snow fall
Infrared radiometer measures the temperature of cloud tops from satellites and creates temperature profiles of the atmosphere from these readings
Microwave radiometer sees through clouds, so provides thorough and accurate atmospheric temperature measurements from satellites

EXTREME WEATHER FACTS

Wettest place Mawsynram, India 467 inches (11,874 mm) in an average year.
Rainiest place Mount Waialeale Crater, Kauai, Hawaii 350 days of rain each year.
Driest place Arica in Chile's Atacama Desert annual average rainfall is one-third of an inch (8 mm).
Highest temperature Al Aziziyah, Libya, recorded a temperature of 136.4°F (58°C) on September 13, 1922.
Lowest temperature Vostok Base, Antarctica, recorded a temperature of –128.6°F (–89.2°C) on July 21, 1983.
Greatest temperature range Verhoyansk, Central Siberia Summer temperatures in this region can reach 98°F (37°C). Winter temperatures can go as low as –90°F (–68°C).
Highest non-tornado wind gust Mount Washington, USA, recorded a wind gust of 231 mph (371 km/h) on April 12 1934.
Highest recorded tornado wind speed Great Plains tornadoes, 310 mph (500 km/h).
Heaviest recorded hailstones Gopalganj, Bangladesh, recorded hailstones that weighed up to 2 pounds 3 ounces (1 kg) on April 14, 1986, killing 92 people.
Hottest region Sahara Desert belt, Dallol, Ethiopia, averaged a daytime temperature of 94°F (34°C) between 1960 and 1966.

MOST SIGNIFICANT GLOBAL WEATHER EVENTS: 20TH CENTURY

Place	Type	Year(s)
India	drought	1900, 1907, 1965–67
China	drought	1907, 1928–30 1936, 1941–42
Soviet Union	drought	1921–22
Sachel	drought	1910-14, 1940–44, 1970–85
China	typhoons	1912, 1992
China	Yangtze River flood	1931
United Kingdom	Great Smog of London	1952
Europe	storm surge	1953
Iran	flood	1954
Japan	Typhoon Vera	1958
Bangladesh	cyclone	1970
North Vietnam	flood	1971
Iran	blizzard	1972
El Niño	current	1982–83
Philippines	Typhoon Thelma	1991
Bangladesh	cyclone	1991
Honduras and Nicaragua	Hurricane Mitch	1998

In January 2004 the National Oceanic and Atmospheric Administration published a list of the top weather, water and climate events of the twentieth century. Meteorologists and hydrologists selected the most notable tornadoes, floods, hurricanes, climate events and other weather phenomena that marked the century. Factors that were taken into consideration included an event's magnitude, meteorological uniqueness, economic impact and death toll. The events are listed in the table at left. Droughts in Asia and Africa were high on the list because of their severe human and environmental toll. The consequences of El Niño events, such as fires, floods and droughts, also featured on the list.

NAMING HURRICANES

Hurricane-like storms are called by different names in different parts of the world. The name hurricane is used for systems that develop over the Atlantic or eastern Pacific oceans. In the western north Pacific and Philippines, these systems are called typhoons while in the Indian and south Pacific oceans, they are called cyclones.

The practice of naming storms has a long history. In the 1800s, hurricanes in the West Indies were named according to the saint's day on which the storm occurred. For example, Hurricane San Felipe struck Puerto Rico on September 13, 1876. Another storm struck Puerto Rico on the same day in 1928, and this storm was named Hurricane San Felipe the Second. Later, forecasters started using latitude–longitude positions to describe hurricanes, but soon realized it was quicker and easier to use short, distinctive names, which are less subject to error than the older, more cumbersome identification methods. These advantages are especially important in exchanging detailed storm information between hundreds of widely scattered stations, coastal bases and ships at sea.

Using women's names became the practice during the Second World War, following the use of a woman's name for a storm in the 1941 novel *Storm* by George R. Stewart.

Since 1953, tropical storms have been named from lists originated by the US National Hurricane Center and now maintained and updated by an international committee of the World Meteorological Organization. The lists featured only women's names until 1979, when men's and women's names were alternated. Six lists are used in rotation. Thus, the 2004 list will be used again in 2010. The name lists have a French, Spanish, Dutch and English flavor because hurricanes affect many nations and are tracked by the weather services of several countries. The letters Q, U, X, and Y are not included because of the scarcity of names beginning with these letters.

The only time that there is a change in the list is if a storm is so deadly or costly that the future use of its name for a different storm would be inappropriate for reasons of sensitivity. If that occurs, the offending name is struck from the list and another name is selected to replace it. Over time, several names have been changed. On the 2002 list, for example, Cristobal has replaced Cesar, Fay has replaced Fran, and Hanna has replaced Hortense. On the 2004 list, Gaston has replaced Georges and Matthew has replaced Mitch. On the 2006 list, Kirk has replaced Keith.

HURRICANE FACTS

WHICH HURRICANE HAS CAUSED THE MOST DEATHS?

The death toll in the Bangladesh cyclone of 1970 has had several estimates, some wildly speculative, but it seems certain that at least 300,000 people died from the associated storm tide in the low-lying deltas.

WHICH HURRICANE HAS CAUSED THE MOST DAMAGE?

The greatest damage caused by a hurricane, as measured in monetary amounts, was Hurricane Andrew (1992), which struck the Bahamas, Florida and Louisiana, USA, causing damage estimated at US $27 billion. However, if one normalizes hurricane damage by inflation, wealth changes and coastal population increases, the worst was the 1926 Great Miami Hurricane. If this storm hit in the mid-1990s, it is estimated that it would have caused damage of over $80 billion in Florida and Alabama.

WHICH IS THE MOST INTENSE HURRICANE ON RECORD?

Typhoon Tip in the northwest Pacific Ocean was measured on October 12, 1979 to have a central pressure of 870 hectopascals and estimated surface sustained winds of 190 miles per hour (304 km/h). Typhoon Nancy in the northwest Pacific region is listed on September 12, 1961 as having estimated maximum sustained winds of 213 miles per hour (341 km/h) with a central pressure of 888 hectopascals.

Hurricane Gilbert's 888-hectopascal lowest pressure (estimated from flight level data) in mid-September 1988 is the most intense (as measured by lowest sea level pressure) for the Atlantic basin. It is almost 20 hectopascals weaker (higher) than the above Typhoon Tip of the northwest Pacific Ocean.

While the central pressures for the northwest Pacific typhoons are the lowest globally, the north Atlantic hurricanes have provided sustained wind speeds possibly comparable to the northwest Pacific. From the best track database, both Hurricane Camille (1969) and Hurricane Allen (1980) had winds estimated at 190 miles per hour (304 km/h). Measurements of such winds are inherently suspect, however, as even the most sophisticated instruments are damaged or completely destroyed when exposed to these speeds.

WHICH HURRICANE LASTED THE LONGEST?

Hurricane/Typhoon John lasted 31 days as it traveled both the northeast and northwest Pacific basins during August and September 1994. It formed in the northeast Pacific, reached hurricane force there, moved across the international dateline and was renamed Typhoon John, and finally curved back across the dateline and renamed Hurricane John again.

Climate facts

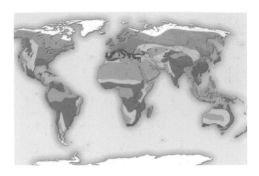

WORLD CLIMATES
The map of global climate zones shows a distribution related primarily to latitude, with regions broadly defined by the tropics of Cancer and Capricorn and the Arctic and Antarctic circles but also affected by altitude and proximity to oceans.

 Tropical　 Mediterranean　☐ Polar

Subtropical　Temperate　Mountain

Arid　Northern　Coastal

Semiarid　temperate

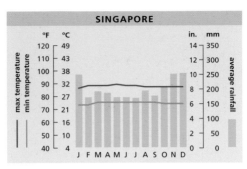

SINGAPORE

■ TROPICAL

Singapore shows little temperature change throughout the year, with the minima and maxima varying by only one or two degrees. Rainfall is high all year, with a peak around November to December.

Tropical climates are found in the equatorial regions that lie between the tropics of Capricorn and Cancer, and feature high temperatures and substantial precipitation throughout much of the year. With little seasonal variation in the intensity of overhead sunlight, temperatures remain high, with the lowest monthly temperature no lower than 65°F (18°C). While there can be rainy and dry seasons, rainfall is typically high, with at least 4 inches (100 mm) of rain every month. Warmth and rainfall provide ideal conditions for life.

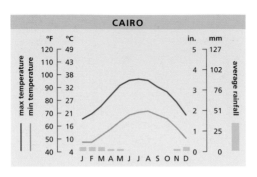

CAIRO

■ ARID

Although rainfall tends to remain low year-round in Cairo, temperatures peak midyear when the Sun is high in a mostly cloud-free sky. Relatively low humidity means large day to night temperature swings.

Arid climates typically create deserts, where the annual precipitation is less than 10 inches (250 mm), and high temperatures ensure that evaporation exceeds this precipitation. Deserts are subject to huge daily temperature fluctuations. The lack of cloud cover allows temperatures to soar during the daytime but fall rapidly after sunset. Many arid zones lie under constantly sinking air such as near subtropical high-pressure cells or downwind of mountains, resulting in cloud-free sky and dry conditions.

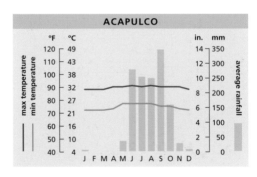

ACAPULCO

■ SUBTROPICAL

Acapulco, Mexico, has elevated temperatures year-round, with a slight cooling in the northern hemisphere winter. Rainfall exhibits a pronounced dry season/wet season cycle, with the peak around June to September.

Found across tropical and subtropical latitudes, the subtropical climate features a distinct wet season/dry season cycle with relatively high temperatures throughout the year. The difference between the warmest and coolest months may amount to only three or four degrees, but the wet season is characterized by high humidities. The seasons of the subtropics are produced by the shifting of high-pressure cells, which move poleward in summer and back toward the equator in winter.

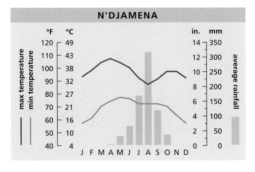

N'DJAMENA

■ SEMIARID

N'Djamena, in the Sahel area of Chad in Africa, is hot throughout the year, but with slightly lower temperatures in July to September. This period corresponds with the wet season, when more than 80 percent of the year's rain occurs.

The semiarid climate zones feature large expanses of grasslands and savannas because the annual precipitation ranges from 10 to 30 inches (250 to 760 mm)—enough water to support some vegetation but too little to sustain full forests. Semiarid regions extend from the tropics into the middle latitudes, wherever passing weather systems supply some moisture. Periods of severe drought also regularly occur. With few trees, these flat, exposed regions are very windy. Seasonal fires and grazing maintain the grasslands.

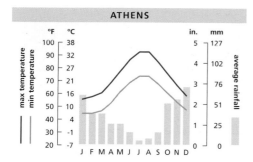

ATHENS

MEDITERRANEAN

The typical Mediterranean climate pattern of hot, dry summers, followed by cool, wet winters, can be seen in this graph of the average monthly temperatures and rainfall experienced in Athens, Greece.

A Mediterranean-type climate is characterized by warm, dry summers and mild, wet winters. It is created by seasonal variations in the position of subtropical high-pressure cells over western sections of the major continents. During summer, these cells drift poleward and their eastern flanks keep the regions dry and warm with sunny skies. In winter, however, the highs drift back toward the equator, permitting rain-bearing midlatitude storms to traverse the regions. Scrublands dominate in Mediterranean climates.

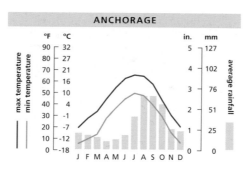

ANCHORAGE

NORTHERN TEMPERATE

Anchorage, in Alaska, USA, experiences long and very cold winters followed by brief, cool summers. The overall climate is quite dry, but with a rainfall peak during late summer to early autumn months.

Since continental landmasses with extreme seasonal temperature contrasts are found primarily in the northern hemisphere, such regions are known as the northern temperate (or boreal) zone. With their large tracts of coniferous forest, they mark a transition between the Arctic tundra to the north and the temperate forest to the south. Typically, the northern temperate zone is marked by a relatively low annual average temperature, with strong seasonal contrasts provided by long, cold winters and short cool summers.

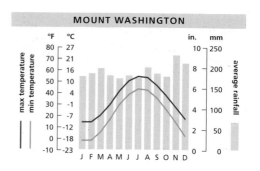

MOUNT WASHINGTON

MOUNTAIN

In the USA, New Hampshire's Mount Washington is 6288 feet (1921 m) high. Winters are cold and windy with frequent snow, but summers are much milder with the average July maxima close to 55°F (13°C).

The effects of altitude and exposure in mountainous areas produce a variety of climate types collectively known as the mountain climate. Mountains intercept and alter moving air masses, creating their own weather patterns. While no single set of characteristics apply to all mountain climates, they differ significantly from nearby valley climates, experiencing lower temperatures, higher winds, greater precipitation, reduced oxygen and greater exposure to ultraviolet sunlight. Latitude determines the altitude at which the zone starts.

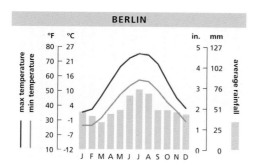

BERLIN

TEMPERATE

The German capital, Berlin, has cold winters, but warm summers in which the maximum temperatures typically reach about 75°F (24°C). Precipitation shows a slight summer peak but is relatively uniform throughout the year.

Temperate climates are found in midlatitudes where nearly half the months have temperatures above 50°F (10°C). These regions all experience four distinct seasons, but the severity of the winter varies according to their proximity to the sea. Along the western edges of the continents, the prevailing ocean winds create a temperate oceanic climate where the lowest monthly temperature rarely falls below 32°F (0°C), with essentially no winter snow cover. Elsewhere, temperate climates may have up to two months of snow.

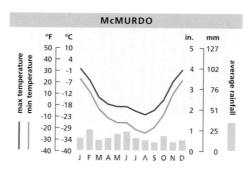

McMURDO

POLAR

Showing the classic profile of extremely cold winters and low precipitation year-round, the polar climate of Antarctica's McMurdo station is harsh. Only highly specialized lifeforms can survive here.

The polar regions of both the northern and southern hemispheres experience extended intervals of darkness and light throughout the year. With long winter darkness and relatively weak summer sunshine, the average temperature of the warmest month typically is less than 50°F (10°C). Precipitation, mostly snow, is also relatively light, with annual totals usually less than 10 inches (250 mm). This harsh climate results in a landscape covered by low-growing tundra vegetation or barren ice cap.

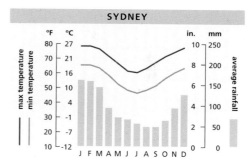

SYDNEY

COASTAL

Sydney, Australia, has a generally mild climate—summer temperatures are moderated by a pronounced sea breeze and rainfall is fairly uniform. The driest period is around the winter–spring months of August and September.

The climate along the shore of an ocean or other large body of water is considerably different from that found even a short distance inland. Because the temperature of the near-surface water changes quite slowly throughout the year, temperature variations along the coast are delayed, producing a relatively stable climate with few fluctuations in temperature. Sea breezes have a moderating effect, lowering temperatures on summer days. In spite of their stable temperatures, coastal regions can be harsh environments.

Billion dollar US weather events 1980–2003

The United States has sustained 57 weather-related disasters over the past 24 years in which overall damages and costs reached or exceeded $1 billion. Seven of these disasters occurred during 1998 alone—the most for any year on record, though other years have recorded higher damage totals.

Two damage figures are given for events before 2002—the first figure represents actual dollar costs at the time of the event and is not adjusted for inflation. Therefore, event costs over time should not be compared using this value. The second value in parenthesis (if given) is the dollar cost normalized to 2002 dollars using a GNP inflation/wealth index. This allows for more accurate comparison of damage figures over time. The total normalized losses for the 57 events are more than $344 billion.

These statistics were taken from a wide variety of sources and represent the estimated total costs of these events—that is, the costs in terms of dollars and lives that would not have been incurred had the event not taken place. Insured and uninsured losses are included in damage estimates, and direct plus indirect deaths—those closely related to the event that would not have occurred otherwise—are included in fatality totals. Economic costs are included for wide-scale, long-lasting events such as drought.

Sources include *Storm Data* (NCDC publication), the National Weather Service, the Federal Emergency Management Agency, other US government agencies, individual state emergency management agencies, state and regional climate centers, and insurance industry estimates.

2003

Hurricane Isabel September 2003. Category 2 hurricane makes landfall in eastern North Carolina, causing considerable storm surge damage along the coasts of NC, VA and MD, with wind damage and some flooding due to 4–12 inch (100–305 mm) rains in NC, VA, MD, DE, WV, NJ, NY and PA; preliminary estimate of over $4 billion in damages/costs; at least 40 deaths.
Severe storms and tornadoes Early May 2003. Numerous tornadoes over the midwest, MS valley, OH/TN valleys and portions of the southeast, with a modern record one-week total of approximately 400 tornadoes reported; over $1.6 billion in damages/costs; 41 deaths.
Storms and hail Early April 2003. Severe storms and large hail over the southern plains and lower MS valley, with Texas hardest hit, and much of the monetary losses due to hail; over $1.2 billion in damages/costs; no deaths reported.

2002

Widespread drought Spring through early fall 2002. Moderate to extreme drought over large portions of 30 states, including the western states, the Great Plains and much of the eastern US; preliminary estimate of over $10 billion in damages/costs; no deaths reported.
Western fire season
Spring through fall 2002. Major fires over 11 western states from the Rockies to the west coast, due to drought and

periodic high winds, with over 7.1 million acres (2.9 million ha) burned; over $2 billion in damages/costs; 21 deaths.

2001

Tropical Storm Allison June 2001. The persistent remnants of Tropical Storm Allison produces rainfall amounts of 30–40 inches (762–1016 mm) in parts of coastal Texas and Louisiana, causing severe flooding especially in the Houston area, then moves slowly northeastward; fatalities and significant damage reported in TX, LA, MS, FL, VA and PA; estimate of approximately $5 (5.1) billion in damage/costs; at least 43 deaths.
Midwest and Ohio Valley hail and tornadoes April 2001. Storms, tornadoes and hail in the states of TX, OK, KS, NE, IA, MO, IL, IN, WI, MI, OH, KY, WV and PA, over a six-day period; over $1.9 billion in damage/costs, with the most significant losses due to hail; at least 3 deaths.

2000

Drought/heatwave Spring–summer 2000. Severe drought and persistent heat over south-central and southeastern states causing significant losses to agriculture and related industries; estimate of over $4 (4.2) billion in damage/costs; estimated 140 deaths nationwide.
Western fire season Spring–summer 2000. Severe fire season in western states due to drought and frequent winds, with nearly 7 million acres (2.9 million ha) burned; estimate of over $2 (2.1) billion in damage/costs (includes fire suppression); no deaths reported.

1999

Hurricane Floyd September 1999. Large category 2 hurricane makes landfall in eastern NC, causing 10–20 inch (254–508 mm) rains in two days, with severe flooding in NC and some flooding in SC, VA, MD, PA, NY, NJ, DE, RI, CT, MA, NH and VT; estimate of at least $6 (6.5) billion damage/costs; 77 deaths.
Eastern drought/heatwave Summer 1999. Very dry summer and high temperatures, mainly in eastern US, with extensive agricultural losses; over $1 (1.1) billion damage/costs; estimated 502 deaths.
Oklahoma–Kansas tornadoes May 1999. Outbreak of F4–F5 tornadoes hit the states of Oklahoma and Kansas, along with Texas and Tennessee, Oklahoma City area hardest hit; over $1.6 (1.7) billion damage/costs; 55 deaths.
Arkansas–Tennessee tornadoes January 1999. Two outbreaks of tornadoes in six-day period strike Arkansas and Tennessee; approximately $1.3 (1.4) billion damage/costs; 17 deaths.

1998

Texas flooding October–November 1998. Severe flooding in southeast Texas from two heavy rain events, with 10–20 inch (254–508 mm) rainfall totals; approximately $1 (1.1) billion damage/costs; 31 deaths.
Hurricane Georges September 1998. Category 2 hurricane strikes Puerto Rico, Florida Keys, and Gulf coasts of Louisiana, Mississippi, Alabama, and Florida panhandle, 15–30 inch (380–762 mm) two-day rain totals in parts of AL/FL; estimated $5.9 (6.6) billion damage/costs; 16 deaths.

Hurricane Bonnie August 1998. Category 3 hurricane strikes eastern North Carolina and Virginia; extensive agricultural damage due to winds and flooding, with 10-inch (254 mm) rains in two days in some locations; approximately $1 (1.1) billion damage/costs; 3 deaths.
Southern drought/heatwave Summer 1998. Severe drought and heatwave from Texas/Oklahoma eastward to the Carolinas; $6–$9 billion (6.6–9.9) damage/costs to agriculture and ranching; at least 200 deaths.
Minnesota severe storms/hail May 1998. Very damaging severe thunderstorms with large hail over wide areas of Minnesota; over $1.5 (1.7) billion damage/costs; 1 death.
Southeast severe weather Winter–spring 1998. Tornadoes and flooding related to El Niño in southeastern states; over $1 (1.1) billion damage/costs; at least 132 deaths.
Northeast ice storm January 1998. Intense ice storm hits Maine, New Hampshire, Vermont, and New York, with extensive forestry losses; over $1.4 (1.5) billion damage/costs; 16 deaths.

1997

Northern Plains flooding April–May 1997. Severe flooding in Dakotas and Minnesota due to heavy spring snowmelt; approximately $3.7 (4.1) billion damage/costs; 11 deaths.
Mississippi and Ohio valleys flooding and tornadoes March 1997. Tornadoes and severe flooding hit the states of AR, MO, MS, TN, IL, IN, KY, OH and WV, with over 10 inches (254 mm) in 24 hours in Louisville; estimated $1 (1.1) billion damage/costs; 67 deaths.
West Coast flooding December 1996–January 1997. Torrential rains (10–40 inches [254–1016 mm] in two weeks) and snowmelt produce severe flooding over parts of California, Washington, Oregon, Idaho, Nevada and Montana; approximately $3 (3.4) billion damage/costs; 36 deaths.

1996

Hurricane Fran September 1996. Category 3 hurricane strikes North Carolina and Virginia; over 10-inch (254 mm) 24-hour rains in some locations and extensive agricultural and other losses; over $5 (5.8) billion damage/costs; 37 deaths.
Southern Plains severe drought Fall 1995 through summer 1996. Severe drought in agricultural regions of southern plains—Texas and Oklahoma most severely affected; approximately $5 (6) billion damage/costs; no deaths reported.
Pacific Northwest severe flooding February 1996. Very heavy, persistent rains (10–30 inches [254–762 mm]) and melting snow over Oregon, Washington, Idaho and western Montana; approximately $1 (1.2) billion damage/costs; 9 deaths.
Blizzard of '96 followed by flooding January 1996. Very heavy snowstorm (1–4 feet [30–122 cm]) over Appalachians, Mid-Atlantic and Northeast; followed by severe flooding in parts of same area due to rain and snowmelt; approximately $3 (3.5) billion damage/costs; 187 deaths.

1995

Hurricane Opal October 1995. Category 3 hurricane strikes Florida panhandle, Alabama, western Georgia, eastern

Tennessee and the western Carolinas, causing storm surge, wind and flooding damage; over $3 (3.6) billion damage/costs; 27 deaths.

Hurricane Marilyn September 1995. Category 2 hurricane devastates US Virgin Islands; estimated $2.1 (2.5) billion damage/costs; 13 deaths.

Texas/Oklahoma/Louisiana/Mississippi severe weather and flooding May 1995. Torrential rains, hail and tornadoes across Texas– Oklahoma and southeast Louisiana–southern Mississippi, with Dallas and New Orleans areas (10–25 inches [254–635 mm] in five days) hardest hit; $5–$6 (6.5–7.1) billion damage/costs; 32 deaths.

California flooding January–March 1995. Frequent winter storms cause 20–70 inches (508–1778 mm) rainfall and periodic flooding across much of California; over $3 (3.6) billion damage/costs; 27 deaths.

1994

Western fire season Summer–fall 1994. Severe fire season in western states due to dry weather; approximately $1 (1.2) billion damage/costs; death toll undetermined.

Texas flooding October 1994. Torrential rain (10–25 inches [254–635 mm] in five days) and thunderstorms cause flooding across much of southeast Texas; approximately $1 (1.2) billion damage/costs; 19 deaths.

Tropical Storm Alberto July 1994. Remnants of slow-moving Alberto brought torrential 10–25 inch (254–635 mm) rains in three days, widespread flooding, and agricultural damage in parts of Georgia, Alabama and panhandle of Florida; approximately $1 (1.2) billion damage/costs; 32 deaths.

Southeast ice storm February 1994. Intense ice storm with extensive damage in parts of TX, OK, AR, LA, MS, AL, TN, GA, SC, NC and VA; approximately $3 (3.7) billion damage/costs; 9 deaths.

1993

California wildfires Fall 1993. Dry weather, high winds and wildfires in southern California; approximately $1 (1.3) billion damage/costs; 4 deaths.

Midwest flooding Summer 1993. Severe, widespread flooding in central US due to persistent heavy rains and thunderstorms; approximately $21 (26.7) billion damage/costs; 48 deaths.

Drought/heatwave Summer 1993. Southeastern US; about $1 (1.3) billion damage/costs to agriculture; at least 16 deaths.

Storm/blizzard March 1993. "Storm of the Century" hits entire eastern seaboard with tornadoes, high winds and heavy snows (2–4 feet [61–122 cm]); $3–$6 (3.8–7.6) billion damage/costs; approximately 270 deaths.

1992

Nor'easter of 1992 December 1992. Slow-moving storm batters northeast US coast, New England hardest hit; $1–$2 (1.3–2.6) billion damage/costs; 19 deaths.

Hurricane Iniki September 1992. Category 4 hurricane hits Hawaiian island of Kauai; about $1.8 (2.4) billion damage/costs; 7 deaths.

Hurricane Andrew August 1992. Category 5 hurricane hits Florida and Louisiana, high winds damage or destroy over 125,000 homes; approximately $27 (35.6) billion damage/costs; 61 deaths.

1991

Oakland firestorm October 1991. Oakland, California, firestorm due to low humidities and high winds; approximately $2.5 (3.5) billion damage/costs; 25 deaths.

Hurricane Bob August 1991. Category 2 hurricane; mainly coastal North Carolina, Long Island, and New England; $1.5 (2.1) billion damage/costs; 18 deaths.

1990

Texas/Oklahoma/Louisiana/Arkansas flooding May 1990. Torrential rains cause flooding along the Trinity, Red and Arkansas rivers in TX, OK, LA and AR; over $1 (1.4) billion damage/costs; 13 deaths.

1989

Hurricane Hugo September 1989. Category 4 hurricane devastates South and North Carolina with 20-foot (6 m) storm surge and severe wind damage after hitting Puerto Rico and the US Virgin Islands; over $9 (13.9) billion damage/costs (about $7.1 [10.9] billion in Carolinas); 86 deaths (57 in US mainland; 29 in US Islands).

Northern Plains drought Summer 1989. Severe summer drought over much of the northern plains with significant losses to agriculture; at least $1 (1.5) billion in damage/costs; no deaths reported.

1988

Drought/heatwave Summer 1988. 1988 drought in central and eastern US with very severe losses to agriculture and related industries; estimated $40.(61.6) billion damage/costs; estimated 5000 to 10,000 deaths (includes heat stress-related).

1986

Southeast drought/heatwave Summer 1986. Severe summer drought in parts of the southeastern US with severe losses to agriculture; $1–$1.5 (1.8–2.6) billion in damage/costs; estimated 100 deaths.

1985

Hurricane Juan October–November 1985. Category 1 hurricane—Louisiana and Southeast US—severe flooding; $1.5 (2.8) billion damage/costs; 63 deaths.

Hurricane Elena August–September 1985. Category 3 hurricane—Florida to Louisiana; $1.3 (2.4) billion damage/costs; 4 deaths.

Florida freeze January 1985. Severe freeze central/northern Florida; about $1.2 (2.2) billion damage to citrus industry; no deaths reported.

1983

Florida freeze December 1983. Severe freeze central/northern Florida; about $2 (4) billion damage to citrus industry; no deaths reported.

Hurricane Alicia August 1983. Category 3 hurricane—Texas; $3 (5.9) billion damage/costs; 21 deaths.

Western storms and flooding 1982–early 1983. Storms and flooding related to El Niño, especially in the states of WA, OR, CA, AZ, NV, ID, UT and MT; approximately $1.1 (2.2) billion in damage/costs; at least 45 deaths.

Gulf States storms and flooding 1982–early 1983. Storms and flooding related to El Niño, especially in the states of TX, AR, LA, MS, AL, GA and FL; approximately $1.1 (2.2) billion in damage/costs; at least 50 deaths.

1980

Drought/heatwave June–September 1980. Central and eastern US; estimated $20 (48.4) billion damage/costs to agriculture and related industries; estimated 10,000 deaths.

TEN MOST DEADLY US TORNADOES

States	Date	Year	Dead	Injured	F-scale
Missouri, Illinois, Indiana	March 18	1925	695	2027	F5
Louisiana, Mississippi	May 7	1840	317	109	Unknown
Missouri, Illinois	May 27	1896	255	1,000	F4
Mississippi	April 5	1936	216	700	F5
Georgia	April 6	1936	203	1600	F4
Texas, Oklahoma, Kansas	April 9	1947	181	970	F5
Louisiana, Mississippi	April 24	1908	143	770	F4
Wisconsin	June 12	1899	117	200	F5
Michigan	June 8	1953	115	844	F5
Texas	May 11	1953	114	597	F5

Weather safety

BE PREPARED

Plan ahead for weather emergencies. Take the time now to choose the best shelter within your home or office. Try to choose a small interior room or stairwell, ideally with walls reinforced with pipes (eg. bathroom) or concrete (eg. basement), on the lowest floor of the building. Make sure everyone concerned knows where to go and what precautions to take (see details below). Also, choose a meeting place where your family can gather after a severe storm, to ensure that you are all safe and accounted for.

Maintain an emergency pack with battery-powered torch and radio, tools for emergency repair, food supplies, first aid, blankets and extra clothing. Keep your car fuel tank full, in case petrol stations close down after a storm. If a severe thunderstorm advice is issued consider precautions such as bringing livestock to shelter, putting away loose objects like garden furniture and parking vehicles under cover. Listen for weather updates and watch the skies. When a warning is issued keep calm, bring children and pets indoors, and close all windows and doors. When the storm approaches be prepared to go to your shelter.

FLOODING

In building If house is in low-lying area be prepared to move to higher ground. Identify where you could go if told to evacuate. Choose several options—a friend's home in another town, a motel or a shelter.
Outside Seek shelter. Do not try to walk through flash floods. Move to higher ground away from rivers, streams, creeks and storm drains.
In vehicle Avoid driving through flooded areas. If caught there, watch for road washouts and avoid dips and underpasses. Do not drive around barricades—they are there for your safety. If your car stalls in rapidly rising waters, abandon it immediately and climb to higher ground. Your life is worth much more than your car.

HAIL

In building Stay away from windows and glass doors. Because large pieces of hail can shatter windows, close your drapes, blinds or window shades to prevent the wind from blowing broken glass inside. Be alert for signs of high winds or tornado (especially if hail is large) and follow tornado precautions if necessary.
Outside Seek cover, face away from wind and protect your head. Be alert for signs of high winds or tornado (especially if hail is large) and follow tornado precautions if necessary.
In vehicle Keep head and face away from windows. Be alert for signs of high winds or tornadoes (especially if hail is large) and follow tornado precautions if necessary.

WIND OR TORNADO

In building Stay inside with doors and windows shut. Stay away from windows, doors and exterior walls. Go to a small, interior room or stairwell on the lowest floor of the building (bathrooms are often best choice). If possible crouch under heavy furniture. Protect your head with cushion or mattress. Interior rooms and halls are the best locations in large buildings. Central stairwells provide good protection, but elevators do not. If the building loses power, you may be in the elevator for a long time. Stay away from glass walls and windows, no matter how small.
Outside Seek shelter in a building (not a car or caravan) immediately. If no suitable shelter is available, lie flat in low dry spot (ravine or ditch) or under a low bridge. Keep alert for fast-moving flash floods. Protect your head. As a last resort, hang on tightly to the base of a shrub or small tree.
In vehicle Do not stay in a vehicle or caravan, and do not try to outrun tornado by driving, especially in populated areas. If possible run to nearby solid structure (shelter or building). If no solid structure is nearby lie flat in dry ditch or ravine outside. Keep alert for flash floods. Protect your head.

LIGHTNING

In building Close windows and doors and keep away from windows, doors and fireplaces. Don't go outside unless it is absolutely necessary. Before storm hits unplug appliances including radio, television and computers and do not touch electrical items or telephones during the storm. Do not take a bath (both water and metal are electrical conductors).
Outside Get inside vehicle or building if possible. Avoid water and objects that conduct electricity (eg. golf clubs, umbrellas, metal fence). Do not stay in open spaces or under tall objects (trees, poles). If no shelter is available crouch down, feet close together with head tucked down. If in a group, spread out, keeping people several yards apart. Unsafe places include underneath canopies, small picnic or rain shelters, or near trees. Where possible, find shelter in a substantial building or in a fully enclosed metal vehicle such as a car, truck or a van with the windows completely shut. Remember, lightning victims can be revived with CPR even though there is no pulse. Injured persons do not carry an electrical charge and can be handled safely.
In vehicle Stay in vehicle with windows closed. Avoid touching metal parts of vehicle. Do not drive, wait. But don't park under trees or other tall objects that may fall over in storm. Be wary of downed power lines that may be touching your car. You should be safe in the car but may receive a shock if you step outside.

Weather organizations

National and international weather organizations observe and predict the weather, working together in a spirit of global cooperation. The national services are designed to respond to the particular needs of the country, and develop expertise in areas of most concern—hurricane forecasting in North America, for example, or rainfall forecasting in Australia. The websites of many weather organizations are a rich source of information and graphics about daily weather as well as storm warnings and weather phenomena. The internet has also enabled amateur weather enthusiasts to share information, experiences and photographs.

International
World Meteorological Organization (WMO)
7 bis, Avenue de la Paix
CP 2300–1211 Geneva 2, Switzerland
www.wmo.ch
Most countries are members. Coordinates World Weather Watch, the World Climate Program, and provides many services, including assistance to developing countries. Publishes a quarterly bulletin.

Australia
Bureau of Meteorology
GPO Box 1289K
Melbourne, Vic 3001
www.bom.gov.au
Head office in Melbourne (open to public) with regional forecasting centers based in other Australian cities. Publishes a range of material.

Australian Meteorological and Oceanographic Society (AMOS)
PO Box 654E
Melbourne, Vic 3001
www.amos.org.au
Mostly professional membership, but open to amateurs. Publishes a bimonthly bulletin. Offices in Hobart, Melbourne and Sydney.

Canada
Atmospheric Environment Service
Environment Canada Inquiry Center
351 St Joseph Bvde, Gatineau
Quebec K1A 0H3
www.ec.gc.ca
National offices in Toronto, Dorval and Hull. Provides information and carries out research. Fact sheets available on topics such as tornadoes, thunderstorms, climate change and the ozone layer.

Canadian Meteorological and Oceanographic Society (CMOS)
PO Box 3211
Station D, Ottawa
Ontario, K1P 6H7

13 centers across Canada hold formal and informal meetings on meteorological and oceanographic subjects. The society publishes a number of journals.

China

China Meteorological Administration
46 Baishiqiao St.
Haidian District, Beijing
www.cma.gov.cn
State-run meteorological service with long association with the WMO. Provides weather information and carries out weather research, including research into disaster prevention.

France

Météo-France
1 quai Branly
75340 Paris Cedex 07
www.meteo.fr
Operates regional and local services for French territory, both in metropolitan France and overseas. A catalog of its many publications is available.

Société météorologique de France (SMF)
1 quai Branly
75340 Paris Cedex 07
Scientific society of both professionals and amateurs. Publishes a quarterly journal.

Germany

Deutsche Meteorologische Gesellschaft (DMG)
Mont Royal, D-56841 Traben-Trarbach
Members include professionals and amateurs. Publishes a magazine.

Deutscher Wetterdienst
Stabsstelle Offentlichkeits-arbeit/Pressesprecher,
Postfach 10 04 65
D-63004 Offenbach
www.dwd.de
Business-related and media forecasting services.

Hong Kong

Hong Kong Observatory
134A Nathan Rd, Kowloon
tel. +852 2926 8200
fax +852 2311 9448
www.hko.gov.hk
Forecasts weather, including hurricanes, and monitors radiation levels in the environment. Publishes a large number of scientific papers.

India

India Meteorological Department
Mausam Bhawan, Lodhi Road
New Delhi 110003
Meteorological and climatic information available. The department's hurricane warning service is one of the oldest in the world and the Research Department holds meteorological records dating back over a century.

Japan

Japan Meteorological Agency (JMA)
Otemachi 1-3-4
Chiyoda-ku
Tokyo 100-8122
www.jma.go.jp
Issues short- and long-range forecasts, including hurricane, tsunami, storm surge, flood and earthquake warnings.

New Zealand

Meteorological Service of New Zealand Limited (MetService)
PO Box 722
Wellington 6015
www.metservice.co.nz
Weather information and forecasting services to media, industry, aviation, agriculture.

Meteorological Society of New Zealand
PO Box 6523, Te Aro, Wellington
Members include professional and amateur enthusiasts. Publishes a quarterly newsletter and biannual journal.

Singapore

Meteorological Services Division
National Environment Agency
PO Box 8, Changi Airport
Singapore 918141
www.app.nea.gov.sg
Provides weather and climatic information for government, business and general public.

South Africa

South African Weather Service
Information and Publication Section
Private Bag X097, Pretoria 0001
www.weathersa.co.za
Operates 16 regional forecasting services, including Marion and Gough Islands in the southern oceans.

United Kingdom

Royal Meteorological Society
104 Oxford Rd, Reading
Berkshire RG1 7LL
www.royal-met-soc.org.uk
Publishes several journals for weather enthusiasts, organizes conferences, and is also involved with educational work. Members include both working scientists and weather enthusiasts in other occupations.

The Met Office
FitzRoy Rd, Exeter,
Devon, EX1 3PB
www.met-office.gov.uk
The UK national weather service. As well as developing computer-based weather forecasting, this office provides information on such vital issues as global climate change and ozone depletion.

European Centre for Medium-Range Weather Forecasts (ECMWF)
Shinfield Park, Reading
Berkshire RG2 9AX
www.ecmwf.int
Provides medium-range forecasts (1 to 10 days) for its 17 member countries and other services around the world. Visits need to be arranged in advance.

United States

American Meteorological Society (AMS)
45 Beacon St, Boston
Massachusetts 02108-3693
www.ametsoc.org
More than 11,000 members, mostly professional meteorologists, oceanographers, and hydrologists. Publishes seven journals, a monthly bulletin, and occasional monographs.

National Climatic Data Center (NCDC)

151 Patton Ave, Asheville
North Carolina 28801-5001
www.ncdc.noaa.gov
Provides climatic data to industry and the general public.

National Oceanic and Atmospheric Administration (NOAA)

National Weather Service
1325 East West Highway
Silver Spring, Maryland 20910
www.nws.noaa.gov
Operates forecasting services nationally.

National Weather Association (NWA)

1697 Capri Way, Charlottesville
Virginia 22911-3534
www.nwas.org
Professional association promoting excellence in meteorology and related activities to an international membership. Publishes a monthly newsletter and a quarterly journal. Anyone with an interest in weather is welcome to join.

Glossary

absolute humidity The mass of water vapor in a unit volume of air.

acid rain An acidic form of rain that occurs when chemicals produced by the burning of fossil fuels mix with water vapor in the air.

aerosonde A small, pilotless aircraft used for recording weather data.

altitude Height above sea level.

alto The prefix used to describe the cloud formations which occur between 6000 feet (2000 m) and 16,500 feet (5000 m).

anemometer A device used to measure the speed of wind.

anvil The top of a thundercloud.

atmosphere The ocean of air that surrounds Earth.

atmospheric pressure Also called air pressure or barometric pressure, the weight of the atmosphere over a unit area of Earth's surface. Changes of weather are usually accompanied by air pressure fluctuations.

aurora A spectacular streak of colored light which occurs when electrically charged particles generated by the Sun strike oxygen and nitrogen molecules in the atmosphere.

Automatic Weather Station (AWS) A weather station which automatically records and transmits data via phone, radio and satellite.

barometer A device used to measure atmospheric pressure.

Beaufort scale A scale devised by William Beaufort in 1805, and used to estimate wind speeds.

blizzard A very severe snowstorm with especially strong winds.

Celsius The unit of measurement used to record temperature in most countries.

chlorofluorocarbons (CFCs) Gases produced by industrial processes. CFCs can reach the upper atmosphere where they interfere with the normal process of ozone layer formation.

cirro The prefix used to describe cloud formations which occur above a height of 16,500 feet (5000 m).

cirrus cloud A white, wispy cloud made of ice crystals and occurring at high levels of the atmosphere.

climate The pattern of weather that occurs in a region over an extended period of time.

coalescence The merging of tiny water droplets in a cloud to form larger drops that may eventually fall as raindrops.

cold front The leading edge of an advancing mass of cold air. When it encounters a mass of less dense warm air, instability results, often triggering heavy rain. Represented on weather maps by a line bearing triangles along one side.

condensation The formation of liquid water from water vapor; occurs when moist air reaches its dewpoint and comes into contact with a solid surface or with condensation nuclei.

convection The upward movement of an air mass warmed by land or sea. As the air rises, condensation may occur, forming clouds.

Coriolis effect The apparent tendency of a freely moving object to follow a curved path in relation to the rotating surface of Earth (to the right in the northern hemisphere and to the left in the southern hemisphere), as shown in the direction of rotation taken by weather systems.

crosswinds Winds blowing at an angle to the direction of movement of a system.

cumuliform cloud A cloud formed by convection that appears flat at the bottom with rounded heaps at the top; the characteristic shape of, for example, cumulus, altocumulus and stratocumulus clouds.

cumulonimbus cloud A cumuliform cloud; the largest in the atmosphere and the producer of thunderstorms. When topped with a large cirrus crown it resembles a giant blacksmith's anvil.

cyclogenisis The formation of a low pressure system in the lower layers of the atmosphere.

desertification The process by which fertile land turns into desert as a result of decreasing rainfall.

dewpoint The temperature to which air must be cooled at constant pressure for saturation to occur, followed by condensation.

diffraction The bending of light when it passes by the edge of an object, resulting in "white" light breaking up into the colors of the spectrum.

downdraft Draft caused by the descent of a cooling column of air. It can result in powerful wind gusts and heavy rain.

dust storm A huge moving cloud of dust.

El Niño A warm current of equatorial water flowing southward down the northwest coast of South America. When pronounced and persistent, it results in rainfall and temperature anomalies over certain areas of the globe.

evaporation The process of changing from a liquid (for example, water) into a gas (for example, water vapor).

eye The clear area of intense low pressure in the center of a hurricane.

Fahrenheit The unit of measurement used to record temperature in the United States.

flash flood A sudden, rapid flood caused by the channeling of a large mass of water through a narrow space such as a canyon or river valley.

fog Cloud that forms in a layer close to, or on, the ground.

fossil The remains, trace or impression of any living thing preserved in, or as, rock.

fossil fuels The remains of organisms or their products embedded in the earth, with high carbon and hydrogen contents. These include fuels such as oil and coal.

front The boundary between two air masses of different temperatures.

frontal system The interface between air masses of different temperatures and humidities, where the most significant weather tends to occur.

glaze A thick coating of clear, smooth ice.

global climate models (GCMs) A computer simulation that reproduces global weather patterns and can be used to predict changes in the weather.

global warming An overall rise in the temperature of Earth's atmosphere.

gravity The force by which bodies are attracted to Earth.

green flash The effect caused as the atmosphere scatters sunlight near sunrise or sunset. Each color of the spectrum becomes briefly visible and the last color is green, because blue, indigo and violet are always blocked by dust.

greenhouse effect The warming caused by certain gases in the atmosphere. These "greenhouse" gases allow sunlight to reach Earth's surface, where it is absorbed and reradiated as heat. The gases then absorb this heat and reradiate it back to Earth.

greenhouse gases Gases that prevent heat escaping from Earth's atmosphere.

Gulf Stream An ocean current that carries warm water from the Caribbean Sea to the North Atlantic Ocean.

hectopascal A unit of measurement used to record air pressure. There are 100 pascals in a hectopascal.

hemisphere One half of Earth. Europe and Northern America are in the northern hemisphere, while Australia and South America are in the southern hemisphere.

high latitudes The regions near Earth's poles, north of the Arctic Circle and south of the Antarctic Circle, which receive the least energy from the Sun and so have cold climates.

high-pressure system An area of high-pressure that rotates clockwise in the northern hemisphere and counterclockwise in the southern hemisphere.

humidity The amount of water vapor in air.

hurricane The term used in North America and the Caribbean for an intense low-pressure cell of tropical origin in which mean wind speeds are greater than 74 miles per hour (118 km/h). Capable of producing widespread wind damage and flooding.

hygrometer A device used to measure humidity.

ice age A cold phase in the climatic history of Earth, during which large areas of land were covered by ice.

instability A condition in which the temperature of a rising air parcel is always warmer than the surrounding air; as a result it continues to rise, sometimes to the troposphere.

interglacial Period between ice ages.

inundation Alternative name for a flood.

iridescence Irregular patches of color in the sky caused by the bending of light around water droplets.

isobar A line drawn on a weather map connecting points of equal air pressure; delineates high- and low-pressure areas. When close together, isobars indicate areas of strong winds.

jet stream Currents of fast-moving air at upper levels of the atmosphere. In middle latitudes, jet streams are more pronounced in winter.

latent heat Heat either released or absorbed when water changes form. Latent heat is absorbed (and the environment is cooled) when water changes from ice to liquid, liquid to vapor, or ice to vapor. Latent heat is released (and the environment is heated) when water changes from vapor to liquid, liquid to ice, or vapor to ice.

latitude A measurement of distance from the equator.

levee An earthen embankment built to block or channel the flow of a river or ocean.

low latitudes The regions nearest the equator, south of the Tropic of Cancer and north of the Tropic of Capricorn, where the Sun is almost directly overhead and its heat is most intense; these regions have hot or warm climates.

low-pressure system A weather system in which air pressure decreases toward the center. This is usually caused by a mass of warm air being forced upward by cold air. Such systems are usually associated with unsettled weather.

meteorology The scientific study of weather.

microclimate A local variation in the normal climate of the region, with differences in temperature and moisture caused by topography, vegetation, or proximity to bodies of water or urban areas.

middle latitudes The regions lying between the Arctic Circle and the Tropic of Cancer (including much of North America and Europe) and between the Antarctic Circle and the Tropic of Capricorn; these regions tend to have moderate climates.

molecule The smallest particle into which a substance can be divided without it becoming something else.

monsoon A seasonal wind which produces heavy rains in tropical and subtropical zones.

occluded front An amalgam of two fronts produced when a cold front catches up with a warm front; usually associated with a low-pressure system.

ocean current A movement of sea water caused by global wind patterns. Ocean currents can carry warm and cold water long distances around the globe.

ozone A gas that absorbs most of the harmful ultraviolet rays from the Sun and also prevents some heat loss from Earth; it occurs naturally in a thin layer in the stratosphere and is also an ingredient in photochemical smog.

ozone layer The thin layer of ozone gas, located roughly 15 miles (24 km) above Earth's surface, which shields us from ultraviolet rays generated by the Sun.

precipitation Moisture in the form of water droplets and ice crystals large enough, and therefore heavy enough, to fall from clouds to Earth.

prevailing winds The most consistent wind patterns for an area

prognostic map A map predicting future weather patterns.

radiosonde A tiny instrument package towed aloft by a balloon. It records air pressure, temperature and humidity.

rain shadow A comparatively dry area on the sheltered side of a mountain range, created by the drying out of air masses as they cross the range.

refraction The change of direction of a light wave when it passes at an angle from one medium to another of different density; refraction causes a light wave to break down into its constituent colors.

relative humidity The amount of moisture in the air at any particular temperature compared to the maximum it could hold at that temperature; usually expressed as a percentage.

rime A white, lumpy coating of ice.

St Elmo's Fire The appearance of a cluster of sparks above a tall object during a thunderstorm, caused by a buildup of electrical charges.

saturation The point at which an air mass can hold no more water vapor at its current temperature; in other words, when the relative humidity of an air mass is 100 percent. At this point, the moisture usually condenses into water droplets. The warmer an air mass is, the more water vapor it can hold.

spectrum The entire range of color that appears to our eye— red, orange, yellow, green, blue, indigo and violet.

squall line Storms occurring simultaneously in a line along a cold front.

storm surge A mound of ocean water drawn up by low pressure below a hurricane; it can cause enormous waves and widespread damage if the hurricane reaches the coast.

stratiform cloud A cloud with a flat, layered appearance; the characteristic shape of, for example, stratus, stratocumulus and altostratus clouds.

stratosphere A layer of Earth's atmosphere extending from the tropopause, about 6 miles (10 km) above the surface of Earth, to the statopause, about 30 miles (50 km) from Earth, just below the mesosphere; the location of the ozone layer.

sublimation The changing of water vapor into ice or the reverse, without it passing through the liquid phase.

supercooled droplets Water droplets cooled to below freezing but still in liquid form.

synoptic forecasting Weather forecasting based on the preparation and analysis of a chart that records surface weather observations taken simultaneously over as wide an area as possible.

synoptic map A chart that shows the weather at a particular time.

temperate Regions which are neither very hot nor very cold. Temperate areas have four distinct seasons.

temperature inversion An atmospheric condition in which cold air at the surface is overlaid by a deck of warmer air. This inhibits vertical mixing of the air masses and often gives rise to smog in urban areas.

thermometer An instrument for measuring temperature.

tornado A spinning column of air that can measure up to 1 mile (1.6 km) in diameter, move at up to 65 miles per hour (105 km/h) and generate winds of up to 300 miles per hour (482 km/h).

tropical cyclone The name given to a hurricane in Australasia and countries around the Indian Ocean.

tropopause An atmospheric transition zone about 6 miles (10 km) above Earth's surface, between the troposphere and the stratosphere, at which temperature stops decreasing with weight.

troposphere The lowest layer of the atmosphere. This is the layer in which we live and in which approximately 99 percent of our weather occurs.

typhoon The name given to a hurricane in the western north Pacific and the China Sea.

updraft Any movement of air away from the ground. The strongest form is found within thunderstorms.

virga Rain that evaporates before it reaches the ground. Virga is often visible as streaks in the sky.

vortex The spinning funnel of a tornado.

warm front The leading edge of an advancing mass of warm air. When it meets a stationary cold air mass, the warm air rises and cools. Condensation may follow, forming clouds, and usually producing widespread precipitation. Represented on weather maps by a line bearing hemispheres along one side.

water vapor Water in the form of gas.

waterspout A spinning column of water-filled air that forms when air currents suck water upward from a lake or ocean.

weather balloon A balloon used to carry meteorological instruments.

weather bureau The place where weather data is collated and analyzed before being supplied to television channels, radio stations and newspapers.

wind shear. Motion caused by one layer of air sliding over another layer that is moving at a different speed and/or in a different direction.

Index